Machine Learning Model Serving Patterns and Best Practices

A definitive guide to deploying, monitoring, and providing accessibility to ML models in production

Md Johirul Islam

BIRMINGHAM—MUMBAI

Machine Learning Model Serving Patterns and Best Practices

Group Product Manager: Ali Abidi
Publishing Product Manager: Ali Abidi
Senior Editor: Tiksha Lad
Technical Editor: Rahul Limbachiya
Copy Editor: Safis Editing
Project Coordinator: Aparna Ravikumar Nair
Proofreader: Safis Editing
Indexer: Pratik Shirodkar
Production Designer: Shyam Sundar Korumilli
Marketing Coordinator: Shifa Ansari

First published: December 2022

Production reference: 2060123

Published by Packt Publishing Ltd.
Livery Place
35 Livery Street
Birmingham
B3 2PB, UK.

ISBN 978-1-80324-990-2

www.packtpub.com

Contributors

About the author

Md Johirul Islam is a data scientist and engineer at Amazon. He has a PhD in computer science and is also an adjunct professor at Purdue University. His expertise is focused on designing explainable, maintainable, and robust data science pipelines, applying software design principles, and helping organizations deploy machine learning models into production at scale.

About the reviewer

Quan V. Dang is a machine learning engineer with experience in various domains, including finance, e-commerce, and logistics. He started his professional career as a researcher at the University of Aizu, where he mainly worked on classical machine learning and evolutionary algorithms. After graduating from university, he shifted his focus to deploying AI products and managing machine learning infrastructure. He is also the founder of the MLOps VN community, with over 5,000 people discussing MLOps and machine learning engineering. In his free time, he often writes technical blogs, hangs out, and goes traveling.

Table of Contents

Part 2: Patterns and Best Practices of Model Serving

3

4

5

6

Batch Model Serving   133

7

Online Learning Model Serving   155

8

Two-Phase Model Serving   179

9

Pipeline Pattern Model Serving   199

10

Ensemble Model Serving Pattern   217

Part 3: Introduction to Tools for Model Serving

14

Part 4: Exploring Cloud Solutions

15

Preface

In this book, we introduce some patterns that should help with serving machine learning models, along with some popular model serving tools. A lot of machine learning projects do not see the light of day due to the technical gap between model creation and model deployment. This book will help machine learning engineers and scientists to get a good understanding of model serving. We have used simple examples and models throughout the book to demonstrate the serving mechanisms instead of building complicated models. This will help you focus on the serving patterns and using the serving tools.

Who this book is for

This book is for data scientists, machine learning engineers, and anyone who wants to work on serving machine learning models.

What this book covers

Chapter 1, *Introducing Model Serving*, introduces model serving and why model serving is important to the success of data science and machine learning projects.

Chapter 2, *Introducing Model Serving Patterns*, describes how patterns in model serving can be of great help to easily identify the best serving approach for a particular problem following the best practices. We also introduce you to different types of serving patterns.

Chapter 3, *Stateless Model Serving*, discusses how stateless model serving can help improve customer experiences, and the advantages of stateless serving in resilient and scalable model serving.

Chapter 4, *Continuous Model Evaluation*, introduces you to continuous model evaluation after serving and why it is important. We also discuss some techniques to evaluate the model continuously.

Chapter 5, *Keyed Prediction*, introduces you to keyed prediction patterns and discusses how passing keys can be beneficial during returning inference to the clients. We also discuss some ideas to generate keys.

Chapter 6, *Batch Model Serving*, discusses batch and offline model serving and how the inference can be updated during batch serving. We also discuss different techniques for updating the model periodically in batch serving.

Chapter 7, *Online Learning Model Serving*, discusses how can we serve models where real-time inferences are needed and some of the techniques and challenges in online serving.

Chapter 8, Two-Phase Model Serving, discusses serving two models in parallel, where one model is strong and the other model is weak. This chapter also discusses the necessity of two-phase serving and some ideas and challenges related to it.

Chapter 9, Pipeline Pattern Model Serving, introduces how models can be served using pipelines using directed acyclic graphs.

Chapter 10, Ensemble Model Serving Pattern, introduces the idea of combining multiple models in serving. It also shows how we can ensemble models in different ways and how the response given to the client is sent as a combined outcome from multiple models.

Chapter 11, Business Logic Pattern, discusses how different business logics are used along with inference codes to serve models.

Chapter 12, Exploring TensorFlow Serving, gives a high level introduction to using TensorFlow Serving tool to serve a model.

Chapter 13, Using Ray Serve, introduces the Ray Serve tool for serving machine learning models with of how to use the tool for serving model following few patterns we have discussed.

Chapter 14, Using BentoML, introduces the BentoML tool for serving models, with examples of using BentoML in ensemble pattern and business logic pattern.

Chapter 15, Serving ML Models Using a. Fully Managed Cloud Solution, discusses how we can serve models using fully managed cloud solution. We use Amazon SageMaker to show you at the high-level how you can serve models using the built-in services provided by a fully managed cloud solution.

To get the most out of this book

This book does not demonstrate creation of the best performing ML models. We only intend to introduce the readers with different model serving patterns and tools and to demonstrate the use of these patterns and tools we have used very basic models. We assume the readers are already familiar with machine learning and know how to create their models. To get best out of the book try touse the patterns and run the examples provided thruoughout the book and use your own models to serve following the patterns discussed.

Software/hardware covered in the book	Operating system requirements
PostMan	Windows, macOS, or Linux
Flask	
TensorFlow	
Ray Serve	
BentoML	
Apache AirFlow	

If you are using the digital version of this book, we advise you to type the code yourself or access the code from the book's GitHub repository (a link is available in the next section). Doing so will help you avoid any potential errors related to the copying and pasting of code.

Download the example code files

You can download the example code files for this book from GitHub at `https://github.com/ PacktPublishing/Machine-Learning-Model-Serving-Patterns-and-Best- Practices`. If there's an update to the code, it will be updated in the GitHub repository.

We also have other code bundles from our rich catalog of books and videos available at `https:// github.com/PacktPublishing/`. Check them out!

Conventions used

There are a number of text conventions used throughout this book.

`Code in text`: Indicates code words in text, database table names, folder names, filenames, file extensions, pathnames, dummy URLs, user input, and Twitter handles. Here is an example: " In this chapter, we will use the `TensorFlow` library."

A block of code is set as follows:

```
train_images = train_images.astype(np.float32) / 255.0
test_images = test_images.astype(np.float32) / 255.0
```

When we wish to draw your attention to a particular part of a code block, the relevant lines or items are set in bold:

```
[[1.5 6.5 7.5 8.5]
[1.5 6.5 7.5 8.5]
[1.5 6.5 7.5 8.5]
[1.5 6.5 7.5 8.5]]
240
```

Bold: Indicates a new term, an important word, or words that you see onscreen. For instance, words in menus or dialog boxes appear in **bold**. Here is an example: "Select **System info** from the **Administration** panel."

Tips or important notes
Appear like this.

Get in touch

Feedback from our readers is always welcome.

General feedback: If you have questions about any aspect of this book, email us at `customercare@packtpub.com` and mention the book title in the subject of your message.

Errata: Although we have taken every care to ensure the accuracy of our content, mistakes do happen. If you have found a mistake in this book, we would be grateful if you would report this to us. Please visit `www.packtpub.com/support/errata` and fill in the form.

Piracy: If you come across any illegal copies of our works in any form on the internet, we would be grateful if you would provide us with the location address or website name. Please contact us at `copyright@packt.com` with a link to the material.

If you are interested in becoming an author: If there is a topic that you have expertise in and you are interested in either writing or contributing to a book, please visit `authors.packtpub.com`.

Share Your Thoughts

Once you've read *Machine Learning Model Serving Patterns and Best Practices*, we'd love to hear your thoughts! Scan the QR code below to go straight to the Amazon review page for this book and share your feedback.

https://packt.link/r/1-803-24990-0

Your review is important to us and the tech community and will help us make sure we're delivering excellent quality content.

Download a free PDF copy of this book

Thanks for purchasing this book!

Do you like to read on the go but are unable to carry your print books everywhere? Is your eBook purchase not compatible with the device of your choice?

Don't worry, now with every Packt book you get a DRM-free PDF version of that book at no cost.

Read anywhere, any place, on any device. Search, copy, and paste code from your favorite technical books directly into your application.

The perks don't stop there, you can get exclusive access to discounts, newsletters, and great free content in your inbox daily

Follow these simple steps to get the benefits:

1. Scan the QR code or visit the link below

https://packt.link/free-ebook/9781803249902

2. Submit your proof of purchase
3. That's it! We'll send your free PDF and other benefits to your email directly

Part 1:
Introduction to Model Serving

In this part, we will give an overview of model serving and explain why it is a challenge. We will also introduce you to a naive approach for serving models and discuss its drawbacks.

This section contains the following chapters:

- *Chapter 1, Introducing Model Serving*
- *Chapter 2, Introducing Model Serving Patterns*

1
Introducing Model Serving

While **machine learning** (**ML**) surprises us every day with new, stunning ideas and demos, a burning question remains: *how can we make the model available to our users?* Often, we see demos of models on different blogs, books, YouTube videos, and so on, and we remain hungry to use the models ourselves. This is where **model serving** comes into the picture. Model serving is how we make our models available for use.

In this chapter, we will learn the definition of model serving, the importance of model serving, the challenges that make model serving difficult, and how people currently serve models, and see some of the available tools used for model serving.

By the end of this chapter, we will understand what model serving is, why model serving is needed, what makes it different from traditional web serving, and how people currently deploy/serve models.

In this chapter, we are going to cover the following main topics:

- What is serving?
- What are models?
- What is model serving?
- Understanding the importance of model serving
- Challenges of serving models
- Using existing tools to serve models

Technical requirements

This chapter does not require you to follow along with any hands-on exercises. However, there are some examples used from the BentoML official site: `https://docs.bentoml.org/en/latest/tutorial.html`.

If you want to try those examples on your local machine, please feel free to install a local version of BentoML following the simple installation instructions here: `https://docs.bentoml.org/en/latest/installation.html`.

Basically, you need to install BentoML like other Python packages using the following command:

```
pip install bentoml
```

Feel free to follow the quick get-started link to understand the steps involved in model serving, which are highlighted in a later section in this chapter.

What is serving?

Serving is an important step for ensuring the business impact of the applications when we develop the life cycle of application development. The application we have developed needs to be available to the user so that they can use it. For example, let's say we have developed a game. After the development, if the game just stays on the developer's machine, then it is not going to be of any use to the users. So, the developer needs to bring the game to the users by serving it through a serving platform such as Apple Store, Google Play Store, and web servers.

So, serving can be seen as a mechanism to distribute our applications/services to end users. The end users can be different based on the applications/services we develop. Serving creates a bridge of communication between the two parties: the developer and the users. This bridge is vital for the business success of our application. If we don't have people using our service, then we are not gaining any business value or impact from the applications we've developed. That's what we have seen in the past when big corporate companies' servers go down for some time: they incur a huge amount of loss. Facebook (Meta) lost ~65 million US dollars due to its outage for some hours in October 2021, as per Forbes: `https://www.forbes.com/sites/abrambrown/2021/10/05/facebook-outage-lost-revenue/?sh=c1d7d03231ad`.

The development-to-serving process usually forms the life cycle of the service or application.

For example, let's consider the life cycle of web application development in *Figure 1.1*.

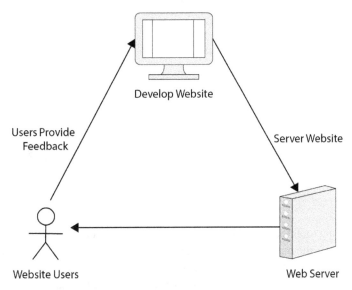

Figure 1.1 – Web development life cycle

The developer develops the website. Then, the website is served using a web server. Only after that are the users able to use the website. The cycle can continue through collecting feedback from users, improving the website, and serving again on the web server.

Now we know what serving is, let's look at what we are actually serving.

What are models?

There are a lot of definitions of **models** from the perspective of various domains. When we define a model or use the term model in this book, we will consider it in the context of ML. A model in ML can be seen as a function that has been fine-tuned through training, using some well-engineered data so that the function can recognize and distinguish patterns in unseen data.

Depending on the problem and business goal, a different model can be used, for example, a linear regression model, a logistic regression model, a naive Bayes model, or a decision tree model. These models' underlying logical representations are different from each other. However, we use the generic term *model* and the problem domain and name of the ML algorithm to give us a clear picture of what the model is, how it was trained, and how the model is represented. For example, if we are told that the model is for linear regression, then we know that it was trained by minimizing a cost function iteratively using the training data, and it can be saved by storing the regression parameters' coefficients and intercepts. Similarly, other models will have different algorithms for training and storing. For a deep learning model, we might have to use forward propagation and backward propagation for training, and for storing we might have to store the weights and biases of all the layers.

The trained model can be stored in different formats to load later for serving and inference. Some popular formats in which to save a model are as follows:

- ONNX

- YAML

- Protobuf

- Pickle

- JSON

- H5

- TFJS

- Joblib

However, model-serving tools usually require the models to be saved in a particular format. So, they provide a function to save the model in its desired format. There are also tools and libraries to convert models from one format to another. For example, in *Figure 1.2*, we see that an AlexNet model that is pre-trained in PyTorch is loaded and exported to ONNX format in a file named `alexnet.onnx`.

> **PyTorch files**
>
> It's worth knowing that PyTorch saves the model using the Python pickle (`https://docs.python.org/3/library/pickle.html`) library. For further reading on PyTorch strategies for saving and loading models, please check out their official documentation: `https://pytorch.org/tutorials/beginner/saving_loading_models.html`.

```python
import torch
import torchvision

dummy_input = torch.randn(10, 3, 224, 224, device="cuda")
model = torchvision.models.alexnet(pretrained=True).cuda()

# Providing input and output names sets the display names for values
# within the model's graph. Setting these does not change the semantics
# of the graph; it is only for readability.
#
# The inputs to the network consist of the flat list of inputs (i.e.
# the values you would pass to the forward() method) followed by the
# flat list of parameters. You can partially specify names, i.e. provide
# a list here shorter than the number of inputs to the model, and we will
# only set that subset of names, starting from the beginning.
input_names = [ "actual_input_1" ] + [ "learned_%d" % i for i in range(16) ]
output_names = [ "output1" ]

torch.onnx.export(model, dummy_input, "alexnet.onnx", verbose=True,
input_names=input_names, output_names=output_names)
```

Figure 1.2 – Example code converting a PyTorch pre-trained AlexNet model to ONNX format

> **Note**
>
> *Figure 1.2* is an example from the PyTorch official website: `https://pytorch.org/docs/stable/onnx.html#example-alexnet-from-pytorch-to-onnx`.

Now we should have a good idea about models and how each model is represented and stored. The following section will introduce us to model serving.

What is model serving?

Like serving a website, we need to serve the trained model so that the model can be used for making predictions to perform business goals. Web/software serving is already at a mature stage. So, we have sophisticated, agreed-upon tools and strategies to serve software. However, ML model serving is still in the phase of growth, and new ideas and tools are coming almost every day.

Model serving can be defined as bringing a model to production by deploying it to a location and providing some access points for users to pass data for prediction and get prediction results.

Model serving usually involves the following steps:

1. **Saving the trained model**: The format in which the model needs to be saved can be different based on the serving tool. So, usually, the serving tools provide a function for saving the model to ensure the model is saved in a format needed by the library.

   Let's use BentoML as an example. We'll cover BentoML in more detail in *Chapter 14*, but in the following code snippet taken from the BentoML official site, `https://docs.bentoml.org/en/latest/tutorial.html`, we see that the popular model-serving library BentoML provides a save function for each of the ML frameworks. During serving using BentoML, we have to call the save method on the appropriate framework. For example, if we have developed a model using `sklearn`, we need to call the `bentoml.sklearn.save_model(<MODEL_NAME>, model)` method to save the model in BentoML format from sklearn format:

   ```
   import bentoml
   from sklearn import svm
   from sklearn import datasets
   # Load training data set
   iris = datasets.load_iris()
   X, y = iris.data, iris.target
   # Train the model
   clf = svm.SVC(gamma='scale')
   clf.fit(X, y)
   # Save model to the BentoML local model store
   ```

```
saved_model = bentoml.sklearn.save_model("iris_clf", clf)
print(f"Model saved: {saved_model}")
# Model saved: Model(tag="iris_clf:zy3dfgxzqkjrlgxi")
```

We can see the list of ML frameworks BentoML currently supports in its GitHub code repository: `https://github.com/bentoml/BentoML`. At the time of writing, they support the following frameworks shown in *Figure 1.3*. We will discuss BentoML in detail in *Chapter 14*.

```
if TYPE_CHECKING:
    from bentoml import h2o
    from bentoml import flax
    from bentoml import onnx
    from bentoml import gluon
    from bentoml import keras
    from bentoml import spacy
    from bentoml import mlflow
    from bentoml import paddle
    from bentoml import easyocr
    from bentoml import pycaret
    from bentoml import pytorch
    from bentoml import sklearn
    from bentoml import xgboost
    from bentoml import catboost
    from bentoml import lightgbm
    from bentoml import onnxmlir
    from bentoml import detectron
    from bentoml import tensorflow
    from bentoml import statsmodels
    from bentoml import torchscript
    from bentoml import transformers
    from bentoml import tensorflow_v1
    from bentoml import picklable_model
    from bentoml import pytorch_lightning
```

Figure 1.3 – BentoML-supported frameworks. Some of these are still in the experimental phase

2. **Annotate the access points**: In this stage, we usually create a service module where we create a function that will be executed when a user makes a request for a prediction. This method is annotated so that after deployment to the model-serving tool, it is exposed via a REST API. BentoML uses a special file, called a `service.py` file, to do this annotation and defining the method that will be annotated. For example, let's look at the `classify(..)` method in the `service.py` file. It has been annotated with `svc.api()` and the input/output formats are also specified. The `service.py` code is annotated with the service access point:

```python
import numpy as np
import bentoml
from bentoml.io import NumpyNdarray

iris_clf_runner = bentoml.sklearn.get( "iris_
clf:latest").to_runner()

svc = bentoml.Service("iris_classifier", runners=[ iris_
clf_runner])

@svc.api(input=NumpyNdarray(), output=NumpyNdarray())
def classify(input_series: np.ndarray) -> np.ndarray:
    result = iris_clf_runner.predict.run( input_series)
    return result
```

3. **Deploy the saved model to a model-serving tool**: In this stage, the model is stored or uploaded to a location needed by the library. Usually, the library takes care of the process behind the scenes and you just need to start the deployment by triggering a command. Sometimes, before deploying a special library, specific packaging might be needed. For example, BentoML creates a special deployable package called a Bento. To build a Bento, you need to first create a `bentofile.yaml` file in the project directory with which to provide the different parameters of the Bento. A sample `bentofile.yaml` file is shown in *Figure 1.4*.

```yaml
# bentofile.yaml
service: "service.py:svc"  # A convention for locating your service: <YOUR_SERVICE_PY>:<YOUR_SERVICE_ANNOTATION>
description: "file: ./README.md"
labels:
    owner: bentoml-team
    stage: demo
include:
- "*.py"  # A pattern for matching which files to include in the bento
python:
  packages:
    - scikit-learn  # Additional libraries to be included in the bento
    - pandas
```

Figure 1.4 – A sample bentofile.yaml file that needs to be created before building a Bento

After that, we can create a Bento using the `bentoml build` command from the command line. The command will build the Bento and you will see some messages in the console, as in *Figure 1.5*.

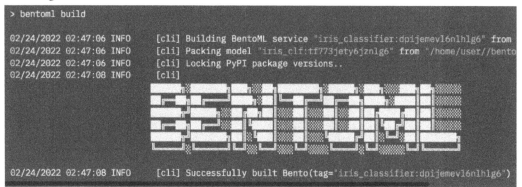

Figure 1.5 – Sample bentoml build output

Please keep in mind that running the `bentoml build` command in a directory with `venv` can take a long time because it scans the whole directory before running the command. This example was run without creating a virtual environment.

Bentos will be saved in a local directory. We can see the Bentos using the `bentoml list` command, as in *Figure 1.6*.

Figure 1.6 – All the Bentos can be seen using the bentoml list command

Then, from the console, we can run the `bentoml serve <MODEL_NAME:TAG> --production` command to serve the model. `<TAG>` can be replaced by the appropriate tag, shown in *Figure 1.6*.

4. **Version controlling of the model**: The model-serving tool also takes care of the version controlling behind the scenes for you. When a new version is uploaded, the APIs exposed from the model-serving tool use the latest model. For example, BentoML uses Tag to refer to different versions. To serve the latest version, you can use `<MODEL_NAME:latest>`. This will pick up the latest-`<MODEL_NAME>`.

In this section, we got a high-level understanding of model serving. In the next section, we will discuss the importance of model serving.

Understanding the importance of model serving

Model serving is one of the critical steps in the ML life cycle but is often neglected. As shown in *Figure 1.7*, users can only start using the model after serving is done. So, model serving is the key step to the business success of a data science or ML team.

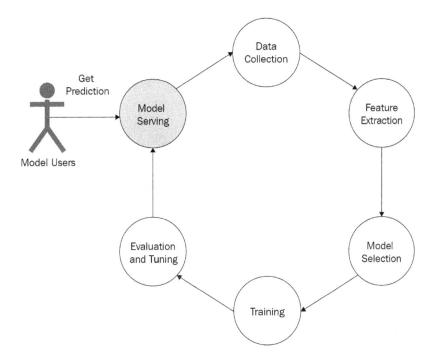

Figure 1.7 – ML life cycle

A lot of models remain unserved simply because model serving is hard. This happens mainly due to the following reasons:

- **Separation of responsibilities**: Often, model development is assigned to data scientists and serving is assigned to engineers, and there may be a gap between the domain knowledge of the two groups. For example, consider a data scientist who has developed a model using a notebook and is getting some predictions and a software engineer who will be serving the model. The product manager is asking for the model to be provided on a server in a production-ready state but some challenges come up:

  - How should they save the model?

  - Are the tools the software engineer is using compatible with the model the data scientist has developed?

- Data scientists tend to use a lot of fancy libraries to make models; are all those libraries supported in the platform the software engineering team is using?

- How will they maintain the version of the model? Is the model developed by data scientists easily maintainable?

- What kind of serving mechanism is needed?

- Which endpoint needs to be published for prediction?

- How does the data from users need to be processed on the server?

Challenges like this create barriers to serving a model after development.

- **Choice of tool**: Tool after tool is appearing for serving models. It makes serving more challenging, as a particular tool may be the perfect choice for a particular type of model, and you need to know which tool is best for which model type. For example, for a simple model, simple REST APIs developed using Flask may be sufficient. However, for complex models, developing a Flask API can be less effective as it was not developed to serve the purpose of stateful client-server communication. Switching from one tool to another can be a problem, and identifying which tool is better for which problem might be challenging. Therefore, as the volume of tools increases, the volume of confusion and challenges in model serving keep increasing.

- **Versioning**: In software engineering, the versioning of software is very easy. We need to redeploy the new updates through a **continuous integration/continuous deployment (CI/CD)** pipeline for a software application. However, in model serving, versioning is challenging because it involves new feature engineering, retraining with new data, and evaluation. Ensuring backward compatibility is not an easy task as it can lead to an error state in ML. Versioning models can be frustrating.

- **Rollback**: In software engineering, let's say deployment to production is behaving incorrectly after the last change; we can then easily roll back the last change and go back to the previous state very quickly. However, in model serving, we cannot take advantage of this shortcut to roll back to the last model.

The preceding points gave us an idea about the challenges involved in model serving. In the following section, we will introduce you to some existing tools for model serving.

Using existing tools to serve models

There are a lot of tools to serve models available now. Some popular tools include the following:

- BentoML (`https://www.bentoml.com/`)

- MLflow (`https://mlflow.org/`)

- KServe (`https://kserve.github.io/website/0.8/`)

- Seldon core (`https://www.seldon.io/solutions/open-source-projects/core`)

- Cortex (`https://www.cortex.dev/`)

- TensorFlow Serving (`https://www.tensorflow.org/tfx/guide/serving`)

- TorchServe (`https://pytorch.org/serve/index.html`)

- Ray Serve (`https://www.ray.io/ray-serve`)

- **Multi Model Server** (**MMS**) (`https://github.com/awslabs/multi-model-server`)

- ForestFlow (`https://github.com/ForestFlow/ForestFlow`)

- DeepDetect (`https://www.deepdetect.com/overview/introduction`)

- Some examples of app-serving tools are CoreML and TensorFlow.js

There are many other tools being developed and made available to users. We will discuss a few of these tools in detail along with examples in the last part of this book.

Sometimes, developers also use very basic REST APIs developed using Flask or FastAPI if the model is simple and does not need frequent updates. This helps software engineers just follow the web development serving life cycle instead of the complex ML model serving life cycle.

These tools aim to reduce the challenges involved in model serving and make resilient model serving easier. However, the availability of a large number of tools also gives rise to confusion in choosing the best tool.

We have now discussed the advantages and challenges of model serving and also introduced some currently available tools for model serving.

Summary

In this chapter, we learned about serving, model serving, and the challenges involved. We learned that model serving is one of the hardest steps in the ML life cycle and, for that reason, is often neglected.

We started our discussion by giving a definition of models and discussing how models are stored. We have seen how models can be stored in a number of formats for serving. However, when using a particular tool for serving, we need to take care to use the format required by that tool.

Then, we discussed model serving. We saw some examples from BentoML, showing the different steps involved. We got an idea of how serving tools can aid you in removing the challenges of model serving.

Then, we discussed the challenges in model serving along with the importance of model serving.

We concluded by introducing you to some existing tools.

In the next chapter, we will introduce you to different model-serving patterns and give a high-level overview of different kinds of patterns we can follow during model serving to make serving resilient and scalable and create a better user experience.

2

Introducing
Model Serving Patterns

Serving a machine learning model is one of the most complex steps in the machine learning life cycle. In *Chapter 1*, we saw why serving machine learning models is challenging. Serving machine learning models involves two groups in two different domains: the ML developer *develops* the model and the software developer *serves* the model. So, we need to agree upon a common language so that we can be sure how our model will be deployed to solve a particular kind of problem. Patterns in design help software architects systematically solve complicated software engineering problems. Similarly, as we learn about design patterns in model serving, the complicated process of model serving will eventually become a piece of cake. This chapter will build on the ideas of some already used patterns for ML serving. We collect the patterns followed by developers and organize and classify those patterns. This chapter will discuss the following topics:

- Design patterns in software engineering
- Understanding the value of model serving patterns
- What are the different model serving patterns
- Examples of serving patterns

Design patterns in software engineering

In the engineering domain, a pattern indicates a common approach or strategy that can be reused. This reuse helps us to understand engineering problems and solve them easily by following the solution pattern that has been made available to us by prior engineers. That's why, when we need to serve a website, we do not have to go back to the theory and try to reinvent the wheel every time. We know the pattern required to serve the web application, which makes our job easier. Most of the time, an engineering team writes down a runbook/docs to solve a recurring problem that appears. This helps engineers avoid debugging the problem every time, thinking of a solution, designing the solution, and applying the solution.

Design patterns are handy to nail hard software engineering problems.

> **The Gang of Four book on design patterns in software engineering**
>
> You might be interested to learn the software engineering design patterns from the book *Design Patterns: Elements of Reusable Object-Oriented Software* by Erich Gamma, Richard Helm, Ralph Helm, and John Vlissides. This book brought about such a dramatic revolution in enhancing the productivity of software development that these four authors became popularly known as the Gang of Four.

To understand how design patterns help us make better software, let's consider a hypothetical problem scenario. We want to make software that will help to create supervised ML models based on customer requirements. It currently supports the following models:

- Linear Regression
- Logistic Regression

A naive solution for this would be the following:

```
class Model:
    def __init__(self, model_name, model_params):
        self.model_name = model_name
        self.model_params = model_params



class ModelTrainer:
    def __init__(self, model):
        self.model= model


    def train(self):
        if self.model.model_name == "LinearRegression":
            trained_model = linear_model.LinearRegression()
https://scikit-learn.org/stable/modules/generated/sklearn.
linear_model.LinearRegression.html
 trained_model.fit(self.model.model_params['X'], self.model.
model_params['Y'])
            return trained_model
        elif self.model.model_name == "LogisticRegression":
            trained_model = linear_model.LogisticRegression()
    trained_model.fit(self.model.model_params['X'], self.model.
model_params['Y'])
```

```
                return trained_model
# Example client calls
model = Model("LinearRegression", {"X": [[1, 1], [0, 0]], "Y":
[1, 0]})
model_trainer = ModelTrainer(model)
trainer_model = model_trainer.train()
```

Now, let's see the problems with this call:

- When a new model needs to be added, ModelTrainer needs to be modified
- It violates the **single responsibility principle**, which says a class should have only a single responsibility, leading to a single reason for the modification
- If we need to add more features for each of the models, we cannot do that easily
- Maintenance becomes difficult for the following reasons: a single method contains all the responsibilities obstructing parallel development, the whole program might break because of a bad change in the single conditional block, and finding the root cause of bugs may be difficult

However, if we look at what this program is doing, it can be seen as a factory providing different trainers (for different models). Then, it becomes easier for us to visualize the problem and also use this common pattern in other similar problems.

Now, let's modify the previous program in the following way:

```
class Model:
    def __init__(self, model_name, model_params):
        self.model_name = model_name
        self.model_params = model_params


class ModelTrainerFactory:
    @classmethod
    def get_trainer(cls, model):
        if model.model_name == "LinearRegression":
            return  LinearRegression Trainer(model)
        elif model.model_name == "LogisticRegression":
            return LogisticRegressionTrainer(model)
class ModelTrainer:
    def __init__(self, model):
        self.model = model
```

```
    def train(self):
        pass


class LinearRegressionTrainer(ModelTrainer):
    def train(self):
      trained_model = linear_model.LinearRegression()
        trained_model.fit(self.model.model_params['X'], self.
model.model_params['Y'])
        return trained_model
# Example client calls
model= Model("LinearRegression", {"X": [[1, 1], [0, 0]], "Y":
[1, 0]})
modelTrainer = ModelTrainerFactory.get_trainer(model)
trained_model = model_trainer.train()
```

The user now gets the desired model trainer from the factory. Whenever a new trainer is needed, we can start providing that trainer from the factory (such as starting a new product from a factory producing products without hampering other production pipelines) by adding a new trainer with a single responsibility. This program is now very easy to maintain and modify. We can use this as a template and apply it in similar problems where we need a collection of different approaches or objects. This pattern is called the **factory pattern**. It is a very basic but useful pattern.

Similarly, there are more than 20 different software engineering patterns that help users to approach common and repeating problems following a well-known template.

In a similar way, using patterns in ML model serving can solve recurring ML model serving problems.

In the next section, we will get a high-level overview of the patterns for serving ML models.

Understanding the value of model serving patterns

Using patterns for ML model serving make us more productive in bringing our model to clients. If we do not follow any patterns, then we may struggle to find the right tool and strategy needed to serve the model for a particular problem.

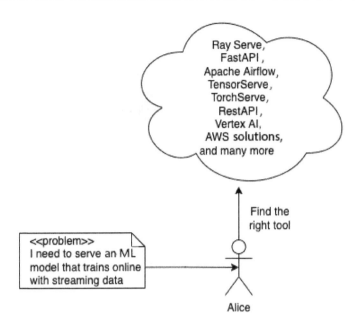

Figure 2.1 – Alice needs to perform trial and error with multiple tools to find the right one

Let's consider the situation of Alice in *Figure 2.1*. Alice has a problem that involves making a data-driven decision. She needs to create a model to solve the problem and deploy the model using a serving tool. She has thousands of tools on offer. She needs to study all these solutions and find the best solution. There is another challenge in the approach of selecting the right tool. Alice is at risk of making a bad choice of tool, as she is solving an optimization problem manually and can be stuck at local maxima. This is always an impediment to productivity, as it involves extra manual effort.

Alice might often have to backtrack to find a suitable model, which creates an exponential search space for her. This situation brings a big tech debt to ML developers because the company might move forward with bad choices of tools that need to be replaced in the near future.

Let's think about it from the point of view of hiring managers. The hiring manager now needs to solve a difficult hiring problem to find suitable talent. They will have difficulties and challenges finding skilled developers who can come up with a solution within a reasonable amount of time. It might be more intuitive to think mathematically about why finding a skilled developer may be hard. Let's say that company A usually faces P kind of problems, each of which needs a different model serving approach. There are N different tools available to serve the model.

So, for each problem, a developer needs to try N tools before finding a satisfactory solution. Therefore, for P problems, there will be P^N different choices for the developer, and in the worst-case scenario, the developer might have to try all these choices to find the best option. Through experience and observations, developers will be able to create a shortlist of the best tools to avoid trying all the choices, and their knowledge of model serving patterns will help the developer to easily make that shortlist. This creates a big bottleneck in productivity. The learning curve to getting skilled in these tools is high. The developer needs to learn the pros and cons of each tool for a particular problem. Therefore, getting a sufficiently skilled developer who can serve the model efficiently becomes hard.

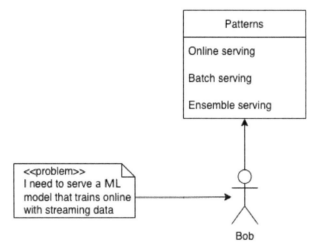

Figure 2.2 – Bob takes the problem pattern and matches it with a few solution patterns

Conversely, let's consider Bob in *Figure 2.2*. As the category of ML problems can be served using only a few recurring patterns, he can quickly map a problem to the serving strategy or pattern needed.

If a problem is encountered, he can quickly map the problem to a suitable serving pattern. Serving becomes a very easy step in the ML life cycle for Bob. Let's revisit the same math problem as before. Now, Bob only has to apply a single pattern for a problem. So, for P problems, he only needs to go deeper into a few patterns. This makes the learning curve easier for a new developer and brings benefits to both the developers and the AI industry.

From this hypothetical scenario, we get the idea that we should follow pattern-oriented approaches in model serving instead of following tool-oriented approaches.

Here are some of the reasons why we need to know model serving patterns:

- **Resilient serving**: The served models should be fault-tolerant and should have mechanisms to recover from faults automatically. However, if we blindly try different approaches and tools without looking at the patterns, we might insert many loopholes, creating issues such as scalability problems and model versioning issues. We might also have difficulty in updating the model without causing problems if patterns are not followed, hampering resilient serving.

- **Using well-vetted solutions**: Using model serving patterns, we can reuse existing solutions to reduce production time. As these approaches are practiced by the developers, these approaches have become trusted and are supported by strong communities. After serving, we can be more confident and also get immediate support if needed.

- **Steep learning curve**: It is obvious that training engineers on model serving becomes much easier when we have patterns ready to hand. The engineers are taught the patterns, and they can follow these patterns to serve a new model without going through an array of tools to select the best one. So, patterns help make the learning curve to get skilled in model serving much smoother.

- **Development of sophisticated serving techniques**: Patterns can help you to develop abstract tools that are specialized for particular patterns. By *abstract tools*, I mean that the tool hides all the implementation details and provides an easier UI interface for users. For example, let's consider the online serving pattern. We can develop abstract tools to serve in real time by simply connecting our stream of data to the tool that already abstracts the background processes and complex logic needed for online learning, as shown in *Figure 2.3*.

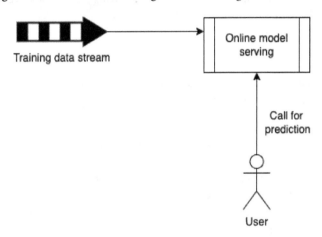

Figure 2.3 – This pattern can help to develop abstract pattern-oriented serving tools

From *Figure 2.3*, we can see that the user is serving models using a tool based on an online model serving pattern. The user does not have to do a lot of complicated tasks such as data cleaning, feature selection, model selection, training, and choosing serving techniques. The user only plugs in the input data to the hypothetical tool and gets the predictions from the APIs exposed to the tool. When we understand the serving patterns for different problem types, developing these tools will be easier.

As Bob is gaining all these advantages from using patterns, he has happy users and a more robust model serving pipeline. On the other hand, as Alice is not following any pattern, she is at high risk of the model experiencing downtime, a sudden drop in performance making clients unhappy, and also dealing with the flaky nature of model inference.

ML serving patterns

There are patterns that are specific to serving ML models. In this section, we will discuss the patterns for ML model serving. We will see the categories of patterns and describe each of the categories separately.

Model serving patterns can be classified into the following two categories at a high level:

- **Serving philosophies**: This group of patterns mainly concerns the principles and best practices you need to be aware of when serving a model – for example, whether the serving should be stateful or stateless, or whether we should evaluate the performance of the model continuously or intermittently

- **Serving approaches**: This kind of pattern gives a clear picture of different serving approaches – for example, how the model will be served in cases of the presence of a large quantity of distributed data, how the model will be served if we need the immediate impact of the most recent data, or how we will serve on edge devices

We will look at these two categories in more detail in the following subsections. We will describe the two categories and see the patterns under each of these categories.

Serving philosophy patterns

In this section, we will learn about the patterns in model serving that describe the state-of-the-art principles in model serving. We place these patterns under the class of patterns for serving philosophies.

These patterns give us ideas about the best practices that we should follow whenever we want to serve models. These patterns, instead of suggesting a particular deployment strategy, provide principles we should follow in all the serving strategies. From these kinds of patterns, we learn to make model serving resilient, available, and consistent, meaning that the responses are the same given the same input. For serving web applications, there are already agreed-upon principles and protocols – for example, communication to a server happens through REST APIs. Similarly, in this section, we will learn some standard principles for ML model serving.

Based on serving philosophies, we can classify serving patterns into three categories, as introduced in the book *Machine Learning Design Patterns* by Michael Munn, Sara Robinson, and Valliappa Lakshmanan:

- Stateless serving

- Continuous model evaluation

- Keyed prediction

We want to avoid *stateful* serving. That is an anti-pattern and should not be classified as a pattern for serving.

Stateless serving

In web serving, the server does not store any **state information, meaning** any client data needed to serve the calls for that particular client (we will go into more detail on states in *Chapter 3, Stateless Model Serving*). The user needs to transfer all the necessary states if they want to use the web service using a REST API. Anyone who needs to access a web service needs to provide the state information needed, and the web service will store that state information in the placeholders to return the desired response after processing. This ensures the scalability of web APIs, as they can be deployed to any server on an on-demand basis.

REST APIs

Representational State Transfer (**REST**) is a set of architectural constraints for designing APIs. For further reading on REST APIs, please read the original thesis by Roy Thomas Fielding (`https://www.ics.uci.edu/~fielding/pubs/dissertation/top.htm`) that introduced REST, and we can also use the following link to learn about it at a high level: `https://www.redhat.com/en/topics/api/what-is-a-rest-api`.

Whenever we want to make our application stateful – or the business logic demands the application needs to be stateful – then we need to be very careful. In web serving, there is a lot of talk about stateless and stateful serving. Both might be used depending on the requirements of the application. In web applications, states in stateful serving mainly refer to the states or status information in the previous calls. However, in model serving, most of the states will come from the states of the model. If the model stores state during serving, then the model might give different results for the same input at different times.

Let's imagine there is a website to check the time at a given location. Whenever a particular user wants to use the application, the user needs to pass the location information along with the API call. Let's also imagine for a moment that the server stores this state information (location). A user from Los Angeles has made a request to the web service to get the current time. The web service got the location information, stored it within its global state, and returned the information. At the same time, if another user from Sydney makes a request to get the time, they might get the wrong time, as the state in the server points to Los Angeles. Therefore, the application becomes buggy and also not scalable.

We can see graphically in *Figure 2.4* that a stateful application that is storing states within the server can cause inconsistent results. A call, **Call 1**, to the server is made to the server, and before it is processed, another call, **Call 2**, is made. **Call 2** will now have access to the states from **Call 1** that the server has stored in its different placeholders or variables. Therefore, there might be inconsistent results in both **Response 1** and **Response 2**.

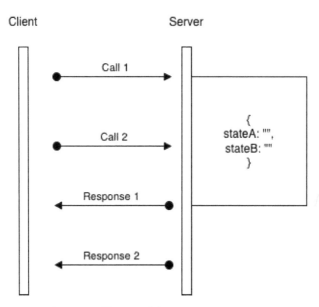

Figure 2.4 – The stateful server stores states

On the contrary, a stateless server requires the client to pass the necessary states. The server is blind to the states of any call and does not store anything related to a call. Each call is served individually and independently. Each of the calls is independent of other calls and does not show any side effect that would result from the mingling of states between the calls.

As shown in *Figure 2.5*, the server does not store any states, and the parallel calls, **Call 1** and **Call 2**, are independent and do not have access to the states of one another.

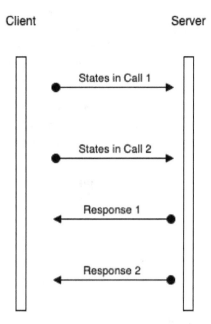

Figure 2.5 – Stateless serving requires the client to pass states

You got the idea of how stateful serving can be problematic. There are many side effects to allowing states within served applications. Some common problems include the following:

- **Scaling the application is difficult**: Whenever we need to deploy the application to a different server, we also need to save the states to the new server. This makes it difficult to scale, and the user might get different results at different times.

- **The results are not consistent**: As the stateful server stores previous states, a new call might use state information from the previous call, giving totally inconsistent results.

- **Security threat**: State information from previous users might be leaked to the new user, which brings the possibility of serious privacy and security threats.

In ML, we have different states that are used during training the model. To export the model for serving using the stateless serving principle, we need to avoid exporting these parameters. Due to the probabilistic nature of the model, sometimes we use different random states during training. While serving the application, we need a mechanism to get rid of this. Otherwise, this will create a bad user experience, as the user will keep getting different results for the same call at different times. Clients might become intolerant of the probabilistic nature of the response from the models, as they are more interested in getting consistent results.

As an example, to understand the problem that stateful model serving might create, let's say a developer has demoed an application to a team manager, and they have seen the result, *R1*. Now, the demo is given to a program manager, and the result is *R2*. This will create distrust and a lack of confidence among the production team as well. Also, if the users start using it and they keep getting different results, *R1*, *R2*, and *R3*, at different times, then it will create a bad customer experience and a high churn rate.

Therefore, we need to use a stateless serving pattern as much as we can. This makes client code and responsibility a little complicated because the client side needs to take a lot of responsibility in extracting the states from the model. However, the hope is that, in the future, more tools will come to remove this client burden.

Sometimes, making stateless serving might be really difficult. For example, let's consider a chatbot application. It needs to store the previous state to make the answers and responses more reasonable.

Continuous model evaluation

One of the key differences between model serving and web serving is that ML models evolve based on data. An ML model will become useless quickly as more and more new example cases appear that aren't taken into account by the model.

For example, let's say that there is a model to detect the house price of a fast-growing city. The model will become stale very soon. Let's say the hypothetical price of a house today is $300,000. After 2 months, the price may become $500,000. So, the model that is developed today cannot be used to make predictions after 2 months. But in a web application, the functionalities of the existing feature do not usually change and the requirements are deployed incrementally step by step. Usually, the deployed requirements do not change significantlly after the **User Acceptance Testing** (**UAT**) is completed.

For example, a feature for user registration might remain exactly the same for years in a web application. However, ML models might need frequent upgrades, as data is growing every day. If we continue to use an old model, then it might suffer from the following problems:

- **Use of stale data**: The model was trained using some data that got stale after a certain period. For example, we can take the use case of house price prediction. Suppose the model was trained with a feature (three bedrooms and two bathrooms) and a target value of $300,000. But the data could become stale after 2 months after which there is a new target value of $500,000 for the same feature. This problem is common in ML. Formally, this problem is known as concept drift and data drift. To learn more about this, please follow this link: `https://analyticsindiamag.com/concept-drift-vs-data-drift-in-machine-learning/`. So, if the model is not updated, the model might be totally useless after a certain period of time.

- **Underfitting**: As new features come in with time, a trained model will behave like an underfitted model in the midst of a new large volume of data. **Underfitting** is a well-known problem that arises if the model is too simple or an insufficient amount of data is used during training. With the new volume of data, an old model will show the impact of underfitting even though it was a very robust model before.

So, we should follow the philosophy of evaluating the model continuously and setting a threshold point at which the model needs to be upgraded.

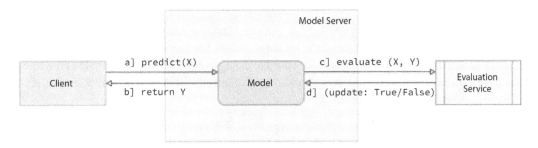

Figure 2.6 – A high-level overview of the continuous evaluation of an ML model

In *Figure 2.6*, we see a high-level overview of serving following the continuous evaluation pattern. Here, we serve the model, and its prediction performance is evaluated using an evaluation service. The evaluation service determines whether we should retrain the model or not based on the performance of the model.

In summary, we need to evaluate the model event after deployment and retrain when it no longer performs well. We can't stay silent after the model is deployed, as the model performance will decay over time.

Keyed prediction model serving

Consider a case where we have a model being served for predicting the house price in a city in the USA. We pass a single input to the model (3 bedrooms and 2 bathrooms), and we get the output of $300,000. This looks very simple for a single-in and single-out case.

However, let's say we send a batch request to the server and the features are provided as an array. Let's assume the input feature array is [(3, 2), (2, 2), (4, 2)]. Now, we get an output of [$300k, $250k, $400k]. Now, if you are asked what kind of house has a price of $250,000, you will answer with the house whose features are (2, 2), meaning it has two bedrooms and two bathrooms. This seems a very fair claim. However, there is a problem here. The answer is assuming the requests are processed sequentially and the response array is filled up sequentially, according to the sequence input features are passed in. Let's pause for a second and think: does this system scale well? If the number of instances in the batch request increases, the time to get a response will keep increasing linearly.

We should rather take advantage of the distributed nature of the servers and parallelize the computation. So, let's say that for our request, the (3, 2) feature has gone to server *S1*, the (2, 2) feature has gone to server *S2*, and the (4, 2) feature has gone to server *S3*. The prediction by server *S3* is completed first, then by server *S2*, and finally, server *S1* completes the prediction for (3, 2). So, the response array is now jumbled, and we get [$400, $250k, $300k].

Now, if we are asked what kind of house has a price of $400,000 and we answer houses with three bedrooms and two bathrooms, our answer will be wrong.

A **keyed prediction** model serving pattern now comes into the picture to solve this problem and enable scalability in the serving of an ML model. The client supplies a **key** along with the features so that the responses can be identified using the key later on. The key can be any value that is distinct and can be used to map to the input instances easily. For example, the key can be as simple as the row number or index of the array element in the input data. The purpose of the key is to be able to match the response against the input instance.

Let's revisit the preceding problem by passing keys now. The request now contains the following instances: `[(k1, 3, 2), (k2, 2, 2), (k3, 4, 2)]`. The response now will be `[(k3, $400), (k2, $250k), (k1, $300k)]`. Therefore, we can easily identify which response belongs to which feature set. Therefore, our problem of leveraging distributed serving is now resolved.

Patterns of serving approaches

In this section, we will discuss the serving patterns that give a clear picture of different serving approaches. These patterns describe which strategy should be followed to serve a particular type of model. These patterns are placed under the classification of patterns of serving approaches.

Serving approaches involve well-vetted strategies to serve an ML model to production – for example, where the model will be served in an online fashion so that the impact of fresh data is immediately visible in the trained model, or in batch mode where the model updates with new training data after some interval. Patterns based on serving approaches categorize different serving strategies.

Some of the main differences between serving philosophy patterns and serving approach patterns include the following:

- A serving philosophy specifies the steps or techniques of serving as fundamental practices. For example, "*during serving, you should log your history*" is a serving philosophy. The serving will still work even if we do not do it – there will be no impact on the serving, but we will be missing a philosophically ideal practice.

- A serving philosophy proposes a set of best practices, but a serving strategy covers a set of *steps*.

- A serving philosophy will not be visible directly, but a serving approach will be – for example, if we serve using a pipeline pattern, we will be able to see the pipeline. However, if we follow a good philosophy of stateless serving, that won't be visible externally.

Based on serving approaches, we can see the following patterns in model serving:

- Batch model serving
- Online learning model serving

- Two-phase model serving
- Pipeline pattern
- Ensemble pattern
- Business logic patter

Batch model serving

Predictions from an ML model are not often possible instantly in a synchronous fashion. Whenever we need a prediction for a single feature set or a small array of feature sets, we might get the response instantly. However, when we need prediction for a large number of instances, we often need to do it asynchronously in a batch manner because of the following reasons:

- The number of instances that need prediction is very high
- The background dataset that needs to be used for training the model is high and the model needs retraining before serving each batch prediction request

For example, let's consider creating monthly sales predictions for different items at different locations of a retail store. For that, we need predictions for thousands of features (locations and items) every month so that the demand planners can make appropriate monthly estimates of sales. For this, the following steps need to take place:

1. The model needs to be updated with large-scale sales data for the last month.
2. The model needs to be redeployed after training with the new data.
3. The model needs to make sales predictions for every feature (location and item).

For example, let's consider the following hypothetical sample prediction of a retail store, X, in Miami, Florida:

- (Miami – apple) -> 20,000 units
- (Miami – swimming goggles) -> 5,000 units

Additionally, for the location in Miami alone, we might need predictions for thousands of items. Considering all the different locations besides Miami, the number of instances needing prediction will be very large, and the prediction needs to happen only after the model is retrained with a new volume of data.

The batch serving pattern deals with this kind of problem, where the model serving solves these batch prediction problems and makes predictions asynchronously in situations when response latency is a big concern.

Online model serving

In online prediction, the model needs to make a prediction immediately after the request is made. Usually, the model makes predictions for a single instance or a small number of instances that can be provided via the HTTP payload limit. In this kind of model, we aim for less latency to provide a better customer experience.

In online models, the model is updated with new features each time a user makes a request using continual learning/online models.

The major advantage of online model serving is that the model stays updated with new data and we can avoid the challenge of retraining.

However, there are some problems that need to be kept in mind while serving the model using the online model serving pattern. Some of these include the following:

- **Model bias**: The model can be converted to a biased model, as the prediction request is available only for a single target class. For example, let's consider we have an online model for deciding on credit applications. If the prediction request comes from only positive cases, then the model might be biased towards predicting positive.

- **Adversarial attacks**: Online models can be susceptible to adversarial attacks. We might need an adversarial detection mechanism before passing the input to the model. Adversarial attacks are also an issue in offline models, but one additional vulnerability of online models is that the attacker can hack the live data ingestion pipeline in online models to provide malicious data.

- If the models are served through multiple servers, then the models will diverge from one another after a short period of time. For example, the model in one server might get a request and get updated, whereas the models in other servers might still stay the same.

Two-phase prediction model serving

In this pattern, two models are served; usually, one model stays on the cloud and the other on the edge devices, but there are other possible differences, including model size or other properties about which we may care. The model on the cloud is often complex and heavy. When handheld edge devices are clients of these models, we have to be aware of the problem that the edge device might be offline or in a weak connection zone. So, we need a lightweight model to be deployed on the edge device to serve the functional requirements within a **Service Level Agreement (SLA)**.

Pipeline model serving

In serving ML, often a complex task is broken down into multiple steps, and each step focuses on a particular ML task. We can structure these steps in the form of a pipeline.

For example, let's imagine we have a computer vision model where we identify different objects from an image or video and provide captions to the objects. The different steps involved in this whole process might be as follows:

1. Preprocess the image.
2. Use an ML model to detect the bounding boxes.
3. Use an ML model to identify the key points.
4. Another Natural Language Processing (NLP) model provides captions based on the key points.

All these separate steps can reside in a separate block in a pipeline. This will give us some flexibility in restarting the pipeline from a failed step whenever needed and debugging the process in a more granular way.

Ensemble model serving

An ensemble pattern comes becomes useful when we need to use predictions from multiple models. This blog gives overviews on ensemble and business logic patterns: `https://www.anyscale.com/blog/considerations-for-deploying-machine-learning-models-in-production`. Some use cases of using ensemble model serving include the following:

- **Updating the model**: After updating the model, we can immediately discard the old model. However, this has some risks. What if the new model does not perform as well as the old model on unseen live data? To avoid this situation, we keep both the old and new models served in an ensemble fashion and then we take the predictions from both models. We keep returning the predictions from the old model during the evaluation period and use an evaluator to evaluate the performance of the new model. Once we are satisfied with the performance of the new model, we can remove the old model.

- **Aggregating the predictions from multiple models**: Often, we need to get predictions from multiple models and aggregate the output either by averaging (for regression) or voting (for classification). This helps us to make more confident predictions.

- **Model selection**: We might have multiple models for different tasks. For example, in the category of vehicle detection, one model may be specialized in detecting cars and another model may be specialized in detecting trucks. So, based on the input feature set, we have to select the right model. Also, the items needing prediction may be from totally different groups. For example, let's say a big retail store has separate models for predicting the price of different groups of items because their feature sets are totally different. The feature set of a grocery item is different from the feature set of a children's toy. So, the company might be interested in having separate models trained for these different types of predictions and the models can remain ensembled, and based on the input feature, we can dynamically select which model to use for prediction.

The preceding scenarios require an ensemble serving pattern, as more than one model is ensembled or stacked together. The serving needs to accommodate this logic to support these problem scenarios.

Business logic pattern model serving

Model serving often requires a lot of business logic to be performed before inference can take place. Any logic that takes place other than inference falls into business logic. Some common business logic includes the following:

- Checking user credentials

- Loading the model from some storage such as S3

- Validating input data

- Preprocessing input data

- Loading precomputed features

These business logic functions require expensive I/O operations. Often, the inference server where the model is served is kept separate from the server where business logic is deployed. Only when the business logic is successful can a user invoke the inference API. For example, confidential military-purpose ML models might not be accessible to any users except those authorized. We need to add business logic to check that authorization. We might have to check for sensitive and malicious data in the input and need business logic to do this. We might have to add business logic to prevent **Distributed Denial of Service (DDoS)** attacks.

Summary

In this chapter, we have learned about the patterns in model serving. We have learned that patterns in model serving can be seen from two angles at a high level: serving patterns based on serving philosophies and serving patterns based on serving strategies.

Serving patterns based on serving philosophies involve the best practices in serving models. These patterns help us ensure resilient model serving by ensuring fault-tolerant, scalable processes in model serving.

Serving patterns based on serving strategies involve recurring approaches used for serving models for different business use cases – for example, a batch serving strategy if the predictions are not necessary immediately and online serving if the predictions are needed immediately.

We also discussed a high-level overview of each of the patterns. We saw that the serving principles such as stateless serving, continued model evaluation, and keyed prediction can help the uninterrupted and resilient serving of the model.

The serving strategy patterns such as batch serving, online serving, two-phase model serving, pipeline patterns, ensemble patterns, and business logic patterns can help us to serve models for different business use cases.

These kinds of patterns can help reduce tech debt in model serving and inspire the future development of pattern-oriented model serving tools.

Further reading

In this section, you can find some further reading that can help you to do further study of the concepts we discussed in this chapter:

- *Machine Learning Design Patterns* by Michael Munn, Sara Robinson, and Valliappa Lakshmanan
- REST APIs: `https://restfulapi.net/`

Part 2: Patterns and Best Practices of Model Serving

In this part, we explore some patterns of serving machine learning models and demonstrate how these patterns can be used through examples.

This part contains the following chapters:

3

Stateless Model Serving

In this chapter, we will talk about **stateless model serving**, the first pattern-based on serving philosophies. We will first talk about stateless and stateful functions to give you an introduction to these concepts. We will see that stateful functions depend on states within the model and also on the states from the previous calls. Due to these strongly coupled dependencies on states, it is difficult to scale the serving.

When serving is desired, the model is served in a stateless manner so that the model does not depend on previous calls, can be scaled easily, and the output is consistent. However, some machine learning models are by default stateful, and we can attempt to reduce the variance of the server model by using some tricks such as specifying random seeds and using hyperparameters that reduce the variance of the model due to states.

This chapter will give an overview of stateful and stateless functions along with examples. Then we will discuss different kinds of states in machine learning models and how those states are not congenial to scalable and resilient serving. We will also introduce some ideas for reducing the impact of those states.

In this chapter, we will discuss the following topics:

- Understanding stateful and stateless functions
- States in machine learning models

Technical requirements

In this chapter, we will go through some programming examples. To run these programming examples, you need to have a Python IDE setup. You can use Anaconda, PyCharm, or any other IDE you prefer:

- Anaconda download link: `https://www.anaconda.com/products/distribution`
- PyCharm download link: `https://www.jetbrains.com/pycharm/download/`

You can see the example code from this chapter and write it manually by yourself, or you can directly clone it from GitHub: `https://github.com/PacktPublishing/Machine-Learning-Model-Serving-Patterns-and-Best-Practices`.

This GitHub repository contains all the code organized by chapters. The code for *Chapter 3* is present here: `https://github.com/PacktPublishing/Machine-Learning-Model-Serving-Patterns-and-Best-Practices/tree/main/Chapter%203`.

Understanding stateful and stateless functions

Before discussing states in machine learning, we should first understand what stateful functions are and how stateful functions can involve difficulties in serving the function to clients. Therefore, let's begin by developing a clear understanding of stateful and stateless functions and the differences between them in this section.

Stateless functions

A stateless function does not have any state within the function that can impact the behavior of the function. The output of the function will be the same if the same input is provided. It will also be independent of the platform where the function is stored.

Stateless functions can behave as pure functions. A pure function has the following properties:

- A pure function's output is identical with identical input. So, $y = f(x)$ is always true. For the same input x, the output is always y. No internal states or variables impact the output and the output is deterministic.

- A pure function will not have any side effects. It will not modify any local variable or state once it is invoked.

For example, consider the following function:

```
def fun(x):
    return x * x
if __name__ == "__main__":
    print(fun(5)) # Always returns 25
```

This function is stateless as its output is always the same if the same input is passed. The output from this function neither depends on the states from the function nor makes any modification to any local states. If we look at the advantages of this function, then we will see that the stateless function has the following benefits:

- The output is consistent. So, in a particular instance, all the clients calling the function will get the same result.

- The function can be served through multiple servers to provide consistent service to millions of customers. As this function does not depend on any local state, it can be easily copied to multiple servers and a load balancer can distribute the traffic to different servers. Traffic pointed to different servers has the same consistent results.

We see stateless functions are very convenient from the perspective of serving. However, sometimes we also need to serve stateful functions. In the following section, we will learn about stateful functions, and how can we extract states and convert stateful functions to stateless functions.

Stateful functions

Stateful functions are functions where the output of the function is not solely defined by the input. Stateful functions have some states hidden *within* the function that have an impact on the output. So, the output is state-dependent and can vary based on the state.

The output of a stateful function depends on the states inside the function and the output will no longer be identical. On the other hand, an invocation of the function can also make modifications to local states. For example, let's consider the following function:

```python
def fun(x):
    import random
    y = round(10*random.random())
    return x + y
if __name__ == "__main__":
    print(fun(5)) # fun can return any number from 5 to 15 for
the same input 5
```

The preceding function is an example of a stateful function where the internal random state, y, impacts the output of the function.

Why is this problematic?

- The function provides different outputs for the same input. The result does not depend solely on client-passed parameters. So, this function can be a cause of frustration as the clients expect consistent output when they provide the same input unless the business logic of the program expects randomized output, as in some games.

- A function with states cannot be scaled very easily. It might not be very clear from this random state why scaling is a challenge with the states of the model. Think about a system that predicts the temperature and uses one mutable internal state that gets mutated by a caller temperature. This service will be hard to scale to a lot of clients and other client requests will be blocked until the current request is finished. However, there are use cases where stateful serving is good. For example, when enhancing models for language translation, we need previous state data.

Now let's look at a function where the result of the current function invocation depends on the previous call, such that the current call of the function uses some states from the last call for producing output.

For example, let's look at the following `counter` function:

```
class Counter:
    count = 0
    def __init__(self):
        pass
    def current_count(self):
        Counter.count += 1
        return Counter.count
if __name__=="__main__":
    counter = Counter()
    print(counter.current_count()) # This call prints 1
    print(counter.current_count()) # This call prints 2
```

In this case, every call to the function depends on the previous call in providing the output.

This dependency causes the following problems:

- Different results based on the order of the calls to the function
- The function cannot be easily served through multiple servers, reducing the scalability options, because the state also needs to be deployed to each server, the servers will easily go out of sync, and the users will keep getting different results from different servers

Extracting states from stateful functions

We have seen two cases where states can cause our program to show inconsistent results and fail to scale easily. The first example has a state within the server that is common to all the functions. The second example shows a scenario where states from previous calls are needed by the current calls.

Now that we know why the preceding functions can be problematic in serving, we need to somehow extract those states or at least reduce the impact of those states.

We can remove the internal state y from the first function and pass the value y from the client side.

The example is shown here:

```
def fun2(x, y):
    return x + y
if __name__ == "__main__":
    import random
    y = round(10*random.random())
    print(fun2(5, y)) # fun2 will return same output for 5 and
same value of y
```

We see the function has been modified from stateful to stateless by extracting the state y and taking the responsibility of passing the state from the client. This kind of conversion adds an extra burden on the client. Now, we notice `fun2` adds the following advantages:

- The output is now consistent. If the client passes the same input data, it will always return the same output data

- The function can be parallelly deployed to many servers and can serve billions of clients without dropping performance

However, with `fun2`, the client has to take responsibility for managing states and passing them to the function.

In web API calls, the client needs to pass all the necessary states with every call. The server does not store anything, indicating each of the calls is independent of one another. That is why these APIs are called **Representational State Transfer** (**REST**) APIs. So, any function that needs to be scaled and resilient needs to be stateless as much as can be.

Similarly, for the second function, we can pass the state count with every call. So, a particular client will always get the same response based on the input the client passes:

```
def current_count(prev_count):
      return prev_count + 1
if __name__=="__main__":
      counter = Counter()
      print(counter.current_count(1))
```

The call to `current_count()` now takes the previous state as shown in the preceding code and removes the dependency on the previous call. So, this call now returns the same output based on the identical input. This can now be scaled easily. The servers will not go out of sync and the users will also get consistent output based on the input they provide.

The preceding examples show some simple stateful functions and how they can be converted to stateless, but when might it be useful to have stateful functions?

Using stateful functions

Stateful does not always mean bad. Sometimes, you need to have some states preserved by the function. For instance, you may be familiar with the **singleton pattern**, which is a famous pattern used widely in software and serving applications. Note that a singleton is a global variable and needs to be used carefully to avoid side effects. For example, privileges for database connectivity for different kinds of users might be different, but if the connection object does not update for different kinds of users, that might cause serious problems. The singleton pattern creates and stores the instance of an object internally as a state and returns the same object in subsequent calls. If the object is already initialized, then it does not re-create the object and returns the already available object. To understand this in practice, let's look at the following code example of the singleton pattern:

```
class DbConnection(object):
    def __new__(cls):
        if not hasattr(cls, 'instance'):
            print("Creating database connection")
            cls.instance = super(DbConnection, cls).__
new__(cls)
            return cls.instance
        else:
            print("Connection is already established!")
            return cls.instance
if __name__=="__main__":
    con1 = DbConnection()
    con2 = DbConnection()
    con3 = DbConnection()
```

This code shows the following output in the console:

```
Creating database connection
Connection is already established!
Connection is already established!
```

From the output, we see that the connection instance is created only once and, based on the state of the connection that is tracked by the function as the internal state, we get a different output. Though, in this case, we get only two different outputs, multiple desirable outputs may be seen based on the state of the function.

Stateful functions can be used in many cases. Some of those use cases may be the following:

- In distributed computing, when we need to keep track of different nodes in the cluster, and to join the results correctly as in **MapReduce**, we need to store the states identifying the nodes and targets during joining.

- In a database, we need to store the indices of the table as states so that we can quickly fetch the data by using the indices.

- During online shopping, the server needs to store the shopping cart information for the user.

- Recommendation engines need to store the customer's previous activities to recommend effective suggestions.

Therefore, stateful functions are not always bad. However, these states must be managed carefully to reduce the side effects of scalable serving. For example, a big data cluster will keep the metadata about all the nodes in some central or master nodes so that other child nodes can scale well. Similarly, we need to be careful in serving stateful applications and need to reduce the impact of serving states as much as possible.

In some models, such as sequence models, stateful serving is the option to choose. These models usually take multiple sequential requests and the inference of one request depends on the previous inference. In this way, keeping the state of the previous stage is required. To learn more about different stateful models, please follow this link: `https://github.com/triton-inference-server/server/blob/main/docs/user_guide/architecture.md#stateful-models`.

In this section, we have discussed stateful functions, and the challenges involved with stateful functions in model serving. We have also seen simple examples of how a stateful function can be converted to stateless. In the next section, we will see some of the common sources of states in machine learning.

States in machine learning models

A machine learning model, at a high level, can be seen as a mathematical function, $y = f(x)$. We provide the input data and train the model. However, the model can perform differently based on the following things:

- **Input data**: The quality of input data, features extracted from the input data, volume of the input data, and so on

- **Hyperparameters**: Learning rate, randomness to avoid bias and overfitting, cost functions, and many more

As these things impact the performance of the model, if they are used as states during serving, scaling can be a challenge, as well as consistency in response to customers.

Now we will look at some cases of how a machine learning model can have states.

Using input data as states

A machine learning algorithm can be designed in such a way that the data used during training is used as states for the model. This can be done in the following two ways:

- Some artificial data is generated to enhance the performance of the model. It is done within the serving of the application and this data can be used for training whenever a retrain action is triggered.

- Input data is used in combination with the features in the feature store (a feature store is a database where we keep the pre-computed features) to update the model in later iterations for online serving. In that case, input from one request is used to alter the model's performance. So, each request is important to the overall model performance.

In both of these cases, the model is dependent on the local states. If we try to scale the model to a second server, we will face difficulty, as these local states need to have consistent values in all the servers. This is almost impossible. We would violate the **CAP principle**.

The CAP principle is a well-known principle in computer science where C stands for **consistency**, A stands for **availability**, and P stands for **partition tolerance**. This principle states that in a distributed system we can guarantee only two of these three properties. To know more about this principle please follow the link here, `https://en.wikipedia.org/wiki/CAP_theorem`.

We can clearly see that if we try to create a replica of the model on multiple servers, we will have obvious issues with consistency. One aspect of the consistency principle is that all the servers should return the same response at the same time from the same request.

However, in this case, we will violate the consistency principle in the following ways:

- Generated data usually involves a stochastic process. So, the data in different nodes will be different, causing the problem of having different models.

- Let's assume a hypothetical scenario, where there is a separate copy of training data in all the nodes that is used to update the models in respective servers. So, there are multiple sources of truth. If the data in one node is modified for some reason, then that node will have a different ground truth for training the model.

- If the data is being read from a streaming source, then the data reading latency in different nodes may be different. So, the data in different nodes will very often fall out of sync. This violates the consistency principle.

- The training convergence in different nodes may take different lengths of time, causing the nodes to go out of sync. This violates the consistency principle.

There can be many other problems if we use the training data as states in a serving model. We might try to solve some of the problems by storing the data files in a central node. This is a good start to resolve some challenges. However, still we are burdened with the following problems:

- The latency of the data being read from that central node by the serving nodes might be different, violating consistency and availability.

- The data communication overhead is high. Packets may be lost, leading to different data at different nodes. If the data is big, then it becomes more challenging. One solution might be to fetch a single batch of data in each call. However, we can clearly see the problem. Thousands of network communication calls are going on with a high payload, incurring costs on our end. Also, as the communications will have different latencies and could also be lost, we are still not solving the fundamental violations of consistency and availability.

We see that using training parameters as state can turn into a big headache during model serving. Other than the problems mentioned previously, there can be many other problems and side effects.

Obviously, the business impact of serving a model with these states will be negative; nobody wants unhappy clients. We should get rid of these states before serving the model.

In this section, we have seen how using training data as states can badly impact serving. In the next section, we will look at random states in a machine learning model.

Random states in model training

Machine learning models are stochastic in nature. So, during the training of the model with the same data, we can have different models with different times giving different predictions.

For example, let's consider the following RandomForestRegressor model:

```
from sklearn.datasets import make_regression
X, y = make_regression(n_features=2, random_state=0,
shuffle=False, n_samples=100)
model = RandomForestRegressor(max_depth=2)
model.fit(X, y)
print(model.predict([[0, 0]]))
```

The output of the model for five runs is the following:

```
[35.67162367]
[31.09473468]
[32.65963333]
[32.29529916]
[28.72626675]
```

We see that the outputs are different at different times even though the same data is being used for training. The output ranges from [28.72626675] to [35.67162367] in just five different runs.

If we look at the tree of the model that is trained in two different runs, we will understand how complex the model is. There are 100 different trees learning in the preceding model.

We can see that number with the following code:

```
print("Total estimators", len(model.estimators_))
```

It will give the following output:

```
Total estimators 100
```

> **Note**
>
> We can also specify the number of estimators during training. We have just used the default parameters to avoid complexity and show how the model can behave differently by selecting the parameters themselves.

To understand how different models can vary, let's visualize only the first tree of the random forest algorithm in two different runs.

Figure 3.1 shows the visualization of the first tree in the first run.

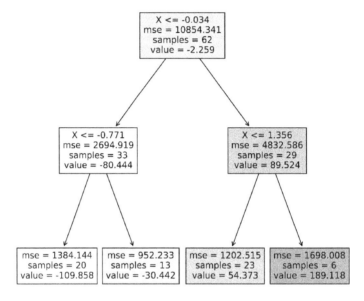

Figure 3.1 – The first decision tree of the random forest learnt in the first run

> **Note**
>
> **Mean Square Error** (**MSE**), shown in *Figure 3.1*, is the average of squares of the differences between predicted and actual values.
>
> For example, let's say for five instances, x_1, x_2, x_3, x_4, x_5, the actual labels are y_1, y_2, y_3, y_4, y_5 and the predicted values from the model are y_1', y_2', y_3', y_4', y_5'.
>
> Therefore, the MSE in this case is $[(y_1 - y_1')\verb|^|2 + (y_2 - y_2')\verb|^|2 + (y_3 - y_3')\verb|^|2 + (y_4 - y_4')\verb|^|2 + (y_5 - y_5')\verb|^|2]/5$.
>
> To read more about MSE, please follow the link: `https://statisticsbyjim.com/regression/mean-squared-error-mse/`.

Figure 3.2 shows the visualization of the first tree in the second run.

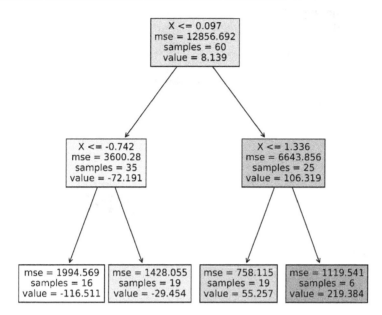

Figure 3.2 – The first decision tree of the random forest in the second run

We have used 100 samples to generate our trained model. If we just look at the root node of both the decision trees in *Figure 3.1* and *Figure 3.2*, we will see many clear differences. The first node in *Figure 3.1* got 62 samples out of 100 after training. The other 38 went to other decision trees that we have not shown. And we see these 62 got further split into two nodes with 33 and 29 samples. The node with 33 samples got further split into two different nodes with 20 and 13 samples. These leaf nodes are also called decision nodes, which return the decision.

Let's see the difference in the two decision trees in the following table only for the first node:

Metric	Tree 1	Tree 2
Condition	X <= -0.034	X <= 0.097
MSE	10854.341	12856.692
Samples	62	60
Value	-2.259	8.139

Figure 3.3 – Differences between metrics on the first decision tree
of the random forests trained with the same data

From the table in *Figure 3.3*, we see the differences are significant. The same data point might have a totally different **decision path** only in the first decision tree.

The decision path is the path followed by the input sample to reach the decision node. For example, if we wanted to predict the output for the input data `Xi = 1.0` using the decision tree in *Figure 3.2*, we would follow this path:

1. Start at the node with condition `X <= 0.097`.

2. Move to the right children node with the condition `X <= 1.336`.

3. As `Xi = 1.0` satisfies the condition `X <= 1.336`, move to the left child with the value = `55.257`.

> **Decision paths**
>
> To learn more about decision paths, you can follow this link: `https://scikit-learn.org/stable/auto_examples/tree/plot_unveil_tree_structure.html#decision-path`.

And if we look at the textual representation of the first decision, at a particular node, we will understand how many states the model is storing just for a single node:

```
DecisionTreeRegressor(ccp_alpha=0.0, criterion='mse', max_
depth=2,max_features='auto', max_leaf_nodes=None, min_
impurity_decrease=0.0, min_impurity_split=None, min_samples_
leaf=1, min_samples_split=2, min_weight_fraction_leaf=0.0,
presort='deprecated',random_state=554039182, splitter='best')
```

All these states can have more or less of an impact on the behavior of the model. If we use some kind of parameter selection algorithm, then the selected parameters and their values might also be different in different runs.

In this section, we have seen that random states within a model can yield inconsistency in responses, causing a bad customer experience. The prediction path might also be different at different times for the same data with the presence of these random states within the model.

In the next section, we will look at the weights and bias states of models, especially deep learning models.

Using model weights as model states

Deep learning models usually have a lot of weights and biases across different layers.

A deep learning model has three kinds of layers: an input layer, hidden layers, and an output layer. There can be one or more hidden layers. Based on the number of hidden layers, a neural network is sometimes referred to as a **shallow neural network** or a **deep neural network**.

To understand the weights and biases in neural networks, we will use the example shown on the official TensorFlow website: `https://www.tensorflow.org/datasets/keras_example`.

We can make some minor changes to the model, as follows:

```
model = tf.keras.models.Sequential([
  tf.keras.layers.Flatten(input_shape=(28, 28)),
  tf.keras.layers.Dense(8, activation='relu'),
  tf.keras.layers.Dropout(0.5),
  tf.keras.layers.Dense(10, activation='softmax')
])
```

The modified code can be found in the GitHub repository for this book, under the folder for Chapter 3. We have changed the hidden layer dimension from 128 to 8 to make it easier to read the states.

Note

Our goal is not to build an efficient machine learning model in any of the chapters. Our goal, rather, is to educate in serving the model, which is the next step after building the model. For further understanding of the importance of the model serving step in the machine learning development life cycle, please refer to *Chapter 1*. Sometimes, we will use some parameters just to ensure better visibility and readability of the states.

We have also changed the epochs from 6 to 3 and added an additional layer, `tf.keras.layers.Dropout(0.5)`, to see the impact of states on different runs. This is an important layer in deep learning models for regularizing and avoiding overfitting.

If we print the model summary after training, using the code `model.summary()`, we will be able to see a summary of the number of parameters or states at different layers. For the example, we extracted data from the official TensorFlow website and modified it a little to see the following output:

```
____
Layer (type)                          Output
Shape                 Param #
=================================================================
===
flatten (Flatten)                     (None,
784)                  0

____
dense (Dense)                         (None,
8)                    6280

____
```

```
dropout (Dropout)                    (None,
8)                        0



____
dense_1 (Dense)                      (None,
10)                       90
=================================================================
===
Total params: 6,370
Trainable params: 6,370
Non-trainable params: 0
```

We see there are 4 layers and in total there are 6,370 parameters. The model we used is very simple and we have also reduced the model size to ensure we can observe the parameters better. Out of those layers, there are parameters only on the two dense layers. The Flatten and the Dropout layers do not have any parameters or states.

Now we can do some mathematics to understand how the number of parameters in the first dense layer is 6,280 and in the second dense layer is 90.

Each of the mnist images we are using is of size (28, 28). So, if we vectorize this (28, 28) matrix, we get a vector of size 28*28 = 784.

Now, each of these 784 pixels will densely connect with the following dense layer, forming connections of size (784, 8) as the dense layer has a dimension of 8. So, in total, there are 784*8=6,278 connections from the input layer to the dense layer. And the first dense layer has 8 nodes. Each node has a bias value adding 8 more parameters for 8 nodes in this dense layer. So, in total, there are 6,278 + 8 = 6,280 parameters for the first dense layer.

In the second dense layer, there are 90 parameters. Let's compute where these 90 parameters come from.

The first dense layer connects to the second dense layer, the output layer. Each of the 8 nodes from the first dense layer connects with each of the 10 nodes in the second dense layer. So, we get a total of 8*10 = 80 connections. Each of these 80 connections has a weight. And each of the 10 nodes in the output dense layer has a bias. So, in total 80+10 = 90 parameters for the last dense layer.

Let's try to see the value of these states. We add the following code after training the model to see the weights and bias of the dense layers after training:

```
W = model.variables
print(len(W))
print("Weights for input Layer to hidden layer")
print(W[0])
print("Shape of weights for input layer to hidden layer", W[0].
```

```
shape)
print("Bias of the Hidden Layer")
print(W[1])
print("Shape of the bias of the hidden layer", W[1].shape)
print("Weights of the hidden to output layer")
print(W[2])
print("Shape of the weights of the hidden to output layer",
W[2].shape)
print("Bias of the output layer")
print(W[3])
print("Shape of the bias of the output layer", W[3].shape)
```

We will be able to see the weights and biases, like the following:

```
Weights from the input layer to the hidden layer
```

Let's look at the output from the following code snippet:

```
print(W[0])
print("Shape of weights for input layer to hidden layer", W[0].
shape)
```

We will see the weights from the input layer to the hidden layer, as follows, along with the information on the dimension of the weights:

```
<tf.Variable 'dense/kernel:0' shape=(784, 8) dtype=float32,
numpy= array([[-0.03122425,  0.08399606,  0.05510005,
...,   0.03683829, -0.08382633,  0.03250936],
```

Here is the truncated output:

```
[-0.08059934,  0.07114483, -0.05455323, ...,
-0.01873366,  -0.08493239, -0.06046978]], dtype=float32)>
```

The shape of the weights from the input layer to the hidden layer is (784, 8).

We see that the size of the matrix of weights in between the input layer and the hidden layer is (784, 8). So, in total, there are at least 784*8 different states that impact the data that proceed from the input layer to the hidden layer during prediction. I am saying *at least* because there are some other states that also impact the manipulation of data, such as bias.

Bias of the hidden layer

We can now observe the output from this part of the above code snippet:

```
print(W[1])
print("Shape of the bias of the hidden layer", W[1].shape)
```

We will print the biases and the shape of the biases of the hidden layer, as follows:

```
<tf.Variable 'dense/bias:0' shape=(8,) dtype=float32, numpy=
array([-0.04289399,
-0.14067535,  0.28432855,  0.17275624,  0.16765627,
0.15334193,  0.14531633, -0.17754607], dtype=float32)>
```

The shape of the bias of the hidden layer is `(8,)`.

For each of the nodes in the hidden layer, there is a bias term. This bias term impacts the data that flows from a node. For example, if the bias value for a particular node is —1 and the data that comes from the input layer to this node is 1, then the data that will flow out from this node is –1 + 1 = 0. That means this node will not have any contribution in the following layer.

Weights from the hidden to the output layer

To get the weights from the hidden layer to the output layer, we use the following code snippet:

```
print(W[2])
print("Shape of the weights of the hidden to output layer",
W[2].shape)
```

This produces the weights and the shape as follows:

```
<tf.Variable 'dense_1/kernel:0' shape=(8, 10) dtype=float32,
numpy=
array([[-0.5102042 , -0.29024327,  0.73903835,  0.06289692,
-0.2546525 ,
```

Here's the truncated output:

```
0.462961, -0.7462656 , -1.1991613, 0.40228426,0.14731914]],
dtype=float32)>
```

The shape of the weights matrix from the hidden layer to the output layer is `(8,10)`. Therefore, there are 8*10 = 80 state values that impact the data that will flow to the output layer.

Bias of the output layer

We can get the biases and the shape of the biases of the output layer with the following code snippet:

```
print(W[3])
print("Shape of the bias of the output layer", W[3].shape)
```

The output we get is the following:

```
<tf.Variable 'dense_1/bias:0' shape=(10,) dtype=float32,
numpy= array([ 0.02194504,  0.48662037, -0.3252962
,          -0.07106664, -0.03191056, 0.29797274,
-0.08240528,             -0.35948208,  0.4130473 ,
-0.41163027],dtype=float32)>
```

The shape of the bias of the output layer is `(10,)`.

For each of the 10 output nodes, we have a bias node.

From the states in the preceding code, we can see that in a deep learning model, there can be numerous states that impact the inference from the model. If these states are not carefully managed, they can create an impediment to scalable and resilient serving. To understand why states can cause an impediment to serving, let's look at the following example.

The weights and bias are updated via training. We will see that these outputs and biases are different if we run the training again with the same data. For example, let's look at the bias output of the `dense_1` layer in a different run to see the difference:

```
array([ 0.0370706 , -0.35546538, -0.4211915 ,   0.49760145,
-0.12625118, 0.21710081, -0.28547847,   0.3222543
,          -0.26602563,  0.27582088], dtype=float32)>
```

The shape of the bias of the output layer is `(10,)`.

We see each of the bias values for the 10 nodes is different for the two consecutive runs with the same data.

Node	Training 1	Training 2	Difference % = 100*(Training 2 – Training 1)/Training 1
1	0.02194504	0.0370706	68.92%
2	0.48662037	-0.35546538	-173.05%
3	-0.3252962	-0.4211915	29.48%
4	-0.07106664	0.49760145	-800.19%
5	-0.03191056	-0.12625118	295.64%
6	0.29797274	0.21710081	-27.14%
7	-0.08240528	-0.28547847	246.43%
8	-0.35948208	0.3222543	-189.64%
9	0.4130473	-0.26602563	-164.41%
10	-0.41163027	0.27582088	-167.01%

Figure 3.4 – Differences of biases in the last dense layer in two consecutive runs

From the table in *Figure 3.4*, we see that the value of weights and bias weights vary a lot in two consecutive runs. While training deep neural networks random initialization of the weights and biases is needed for the successful convergence of the model. However, this creates an additional burden for us while serving the model. These states have an impact on the model output. So, the model that is trained does not behave as a pure function and can have different values, depending on the values of these states.

We can save the model to a directory called `saved_model` using the following command:

```
model.save('saved_model')
```

After the model is saved, we will see a structure like the following:

```
saved_model
     -> assets
     -> variables
 -> variables.data-00000-of-00001
 -> variables.index
     -> saved_model.pb
```

We notice the model structure is located in the `saved_model.pb` file and all other states are stored in the assets and variables folder. We can also save the model as JSON by converting the model to JSON using the following command:

```
json = model.to_json()
```

We can then view the model structure as follows in *Figure 3.5*:

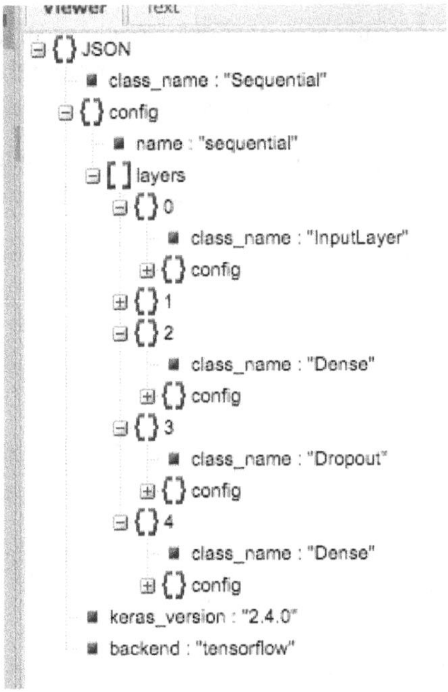

Figure 3.5 – JSON view of the JSON representation of the model

This view is generated from the public tool at `http://jsonviewer.stack.hu/`. You need to paste the JSON there and it will create a view like the one in *Figure 3.5*.

With all these states, this model is heavily dependent on the local environment.

This stateful nature of the model due to the states from weights and biases will create the following problems in model serving:

- If we serve a different copy of the model using a different pipeline, then the response from the models will be different even though the models are following the same algorithm and the same input data distribution. So, a client making a call to the prediction API might get two different responses at two different times due to a load balancer forwarding the requests at different times based on traffic loads.

- Scaling the serving is difficult as the models might go out of sync for reasons related to retraining, transfer learning, online learning, and many other reasons.

- For continuous evaluation, the models on different servers will need to update at different times and will quickly go out of sync. For example, let's say we monitor the performance of three copies of the same model on three servers: S1, S2, and S3. We use the same metric that we used during training to evaluate the performance of the models. Let's say the performance of the model on S1 falls below the threshold value of the accepted error rate (which depends on the problem and business goal) at time T1. In this case, we need to do some hard work. We have to retrain the model and copy the reatried model to all the single servers before the responses from all the servers can be in sync. If we just retrain the model on server S1, then the model will be out of sync with the other ones.

- During online training, it is hard to train the models continuously, as whenever new data comes in, we need to retrain and the models on all the servers will become out of sync very quickly.

In this section, we have seen how we can compute the number of parameters in a basic densely connected neural network. In the following section, we will see the number of parameters in a Recurrent Neural Network (RNN).

States in the RNN model

To demonstrate the weights and parameters in the RNN model, let's use the example from the official TensorFlow site: https://www.tensorflow.org/guide/keras/rnn.

The example code is also present in the GitHub repository of this book, under the Chapter 3 folder.

Let's first build the model using the layers shown here:

```
model = Sequential()
model.add(Embedding(input_dim=1000, output_dim=64))
model.add(LSTM(128))
model.add(Dense(10))
```

We will be able to see the number of parameters in different layers using model.summary(). We will get the following model summary after running the model.summary() statement:

```
Model: "sequential"

_____
Layer (type)                    Output
Shape                 Param #
=================================================================
```

```
===
embedding (Embedding)              (None, None,
64)                      64000


___
lstm (LSTM)                        (None,
128)                     98816


___
dense (Dense)                      (None,
10)                      1290
=================================================================
===
Total params: 164,106
Trainable params: 164,106
Non-trainable params: 0
```

Now, let's try to understand the states in each of the layers. The number of parameters in the Embedding layer is found by multiplying the input dimension by the output dimension. That means each of the input edge weights connects to each of the output edge weights in a cross-multiplication manner. So, we get the total parameters in this layer as 1000*64 = 64000.

In the LSTM layer, the number of parameters is 98,816. To compute the number of parameters in the LSTM layer, we need to take the following states:

1. Compute the number of edges going from the input to different units for each of the gates and the number of bias terms for each of the units in the LSTM gate. In LSTM, there are four gates. So, for four gates, we get `4*(input_dim*units+ units) = 4*(64*128+128) = 33,280`.

2. Compute the number of weights that are coming from feedback edges for each output node of the output layer. We know the output dimensions at time T are fed back to each of the units in time T + 1 for recurrency. So, the number of parameters for this recurrency for all four gates is `4*units*units = 4*128*128 = 65, 536`.

3. Add the numbers from *step 1* and *step 2* to get the total number of parameters in the LSTM layer. So, the parameter total for this model is $33, 280 + 65, 536 = 98, 816$.

We have already shown how to compute the number of parameters in the Dense layer in the previous section. Using the same approach in this Dense layer, the number of parameters is = 128*10 + 10 = 1,290.

In the last two sections, we have seen that a deep learning model can have millions of parameters as states within the saved model. These states make the model a **non-pure function** and the response is heavily dependent on the states, which violates a number of fundamental principles in serving, carrying the risk of an unhappy client experience. In the next section, we will look at the parameters or states in a regression model and understand their impact on serving.

States in a regression model

States in a regression model mainly come from the regression parameters. The regression model has two main kinds of parameters. The structure of a regression equation is y = ax + b. Here, the two parameters are a and b:

- **Intercept**: Parameter b in the previous equation is called the intercept. To understand what it implies, let's say x = 0. So, we get y = b. This is a line parallel to the *x* axis and intercepts the *y* axis at point (0, b).

- **Slope**: The a parameter is known as the slope. It can be defined as the rate of change of y or the amount of change of y if x changes by 1. Sometimes this is also known as the coefficient. For each of the input features, there will be a separate coefficient in a simple linear regression model.

To demo the parameters in a simple linear regression model, let us take the following code snippet from `https://scikit-learn.org/stable/modules/generated/sklearn.linear_model.LinearRegression.html`:

```
import numpy as np
from sklearn.linear_model import LinearRegression
X = np.array([[1, 1], [1, 2], [2, 2], [2, 3]])
y = np.dot(X, np.array([1, 2])) + 3
reg = LinearRegression().fit(X, y)
reg.score(X, y)
print(reg.coef_)
print(reg.intercept_)
print(reg.predict(np.array([[3, 5]])))
```

Here, the output we get from the three `print` statements is as follows:

```
[1. 2.]
3.0000000000000018
[16.]
```

We can see there are two slopes/coefficients, 1 and 2, for 2 input features, *x1* and *x2*, in input data *X*. For example, in the first input instance in X, the value of *x1 = 1* and the value of *x2 = 1*. And here, the intercept is *3.0000000000000018~ 3.0*.

So, we can write the regression equation as follows:

$$Y = 1 * x1 + 2 * x2 + 3.0$$

In the third `print` statement, `print (reg.preditc(np.array([3, 5])))`, we are doing a prediction for `[3, 5]` and we have seen the output of 16 as shown in the preceding output, `[16.]`.

Now, let's compute the prediction using the regression formula:

$$Y = 1 * 3 + 2 * 5 + 3 = 3 + 10 + 3 = 16.0$$

We got the same result. From this, we understand these states have a direct impact on the result that we get. Therefore, these states can cause difficulties during serving if not managed properly. For example, if we serve a regression model in different nodes in different regions, those servers can go out of sync if the training data is not updated simultaneously on all servers. This will create difficulties in maintenance and the serving or deployment pipeline can become unstable due to the heavy maintenance overhead.

States in a decision tree model

To understand the states in a decision tree model, let's consider the following code snippet:

```
from sklearn import tree
import matplotlib.pyplot as plt
X = [[0, 0, 2], [1, 1, 3], [3, 1, 3]]
Y = [0, 1, 2]
clf = tree.DecisionTreeClassifier()
clf = clf.fit(X, Y)
print(clf.predict([[3, 1, 6]]))
print(clf.predict([[1, 1, 6]]))
print(clf.decision_path([[3, 1, 6], [1, 1, 6]]))
tree.plot_tree(clf)
plt.show()
```

This is a very basic model to show the states that can affect decision-making. We can see the output tree in *Figure 3.6*.

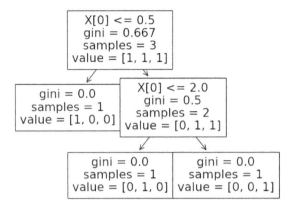

Figure 3.6 – Graphical representation of the decision tree

> **Gini**
>
> In *Figure 3.6*, we can see a metric called **gini**. It is known as the **gini index** or the **gini coefficient**. Gini impurity is the default criteria used by a decision tree in `scikit-learn` to measure the quality of the split of data points to children. If the gini index value is 0, then the samples are assumed to have perfect equality and no further splitting is needed.
>
> That's why we see in all the child nodes in *Figure 3.6* that the gini index is 0. A gini index of 1.0 indicates maximum impurity.
>
> Therefore, as the decision tree evolves from the root node, the gini index decreases, indicating the samples are getting more and more homogeneous.
>
> To read more about the gini index and the formula used to compute the gini index, you can follow this link: `https://en.wikipedia.org/wiki/Gini_coefficient`.

Now let's try to understand how states impact making decisions. During decision-making, the states that will be used are the following:

- Conditional states in each of the non-leaf nodes, such as X[0] <= 0.5

- The value state in the leaf nodes or decision nodes

The state value represents the target label for this node. For example, a leaf node with the value [0, 1, 0] means any instance that comes to this node will be labeled as Y[1] as position 1 is set in the value. Please note that the value has only three elements as our input used during training is of size 3. Now let's try to understand how we can predict X1= [3, 1, 6] using the decision tree shown in *Figure 3.6*:

1. At the root node, we check X[0] <= 0.5. In other words, 3 <= 0.5. As this condition is *false*, we go to the right child node.

2. At the right child node, the condition is X[0] <= 2.0, or 3 <= 2.0. This condition is *false*. So, we reach the rightmost leaf node from here. Here, we see the value is [0, 0, 1].

3. We return the prediction as Y[2] = 2.

We notice that we get the same output from the statement print(clf.predict([[3, 1, 6]])).

If we follow the same steps for the input X2 = [1, 1, 6], we will end up at the leaf node with the value [0, 1, 0], yielding the prediction Y[1] = 1. This is the same result we get from the statement print(clf.predict([[1, 1, 6]])).

We can path followed by the input in the decision tree to make a decision using the decision paths from the statement print(clf.decision_path([[3, 1, 6], [1, 1, 6]])). The output of this statement is the following:

```
(0, 0)  1
(0, 2)  1
(0, 4)  1
(1, 0)  1
(1, 2)  1
(1, 3)  1
```

We see a sequence of tuples (i, j) where i is the sample number in the data passed to the function decision_path(X). For example, in our input *[[3, 1, 6], [1, 1, 6]]*, we have two samples. The sample at index 0 is [3, 1, 6] and the sample at index 1 is [1, 1, 6]. And the term j indicates the node number in the array representation of the tree. During the array representation of a binary tree, the root node is placed at index 0, the left child is placed at index 2*0 + 1 = 1, and the right child is placed at index 2*0 + 2=2. This process continues recursively. The general formula is the left child of a node at index n is placed at index 2*n + 1 and the right child is placed at index 2*n+2.

Other than these two states, conditional expression and value, other states will have an impact on the training time. If those states are different, the models will be different.

In this section, we have seen the states that impact the inference in a decision tree. If we rerun the code shown previously, we will get a different decision tree from *Figure 3.6*. Therefore, these models are very sensitive to states.

Serving this model will encounter similar problems to those listed in earlier sections.

In the last few sections, we have explored states in some basic DNN models, RNN models, ensemble models (the random forest regression model), the linear regression model, and decision tree models. We can use the same strategy to understand different states in other models and to understand how the states impact inference and serving. This understanding of states is a critical step before we can decouple states to ensure scalability and resiliency in model serving.

In the next section, we will discuss how can we remove or reduce the impact of these states during serving.

Mitigating the impact of states from the ML model

We learned about states in some ML models in the last section and have seen how serving can be inefficient in the presence of these states. In this section, we will look at some ideas we can use to minimize the impact of states and ensure resilient serving.

Using a fixed random seed during training

Some models use random seeds during training that are used to initialize the values for some of the states:

1. As an example, let us take the random forest model shown before and add a fixed random state, `random_state=0`, to the model as follows:

    ```
    model = RandomForestRegressor(max_depth=2, n_
    estimators=10, random_state=0)
    ```

    Now, the model will give the same result for the prediction `print(model.predict([[0, 0]]))`.

2. We run the model five times and we get the same output every time:

    ```
    [38.23613716]
    [38.23613716]
    [38.23613716]
    [38.23613716]
    [38.23613716]
    ```

If we use the same data for training, then we will have the same model each time we train unless we don't change the value of `random_state`.

However, if we change the data, then the models will be different.

Moving the states to a separate location

After training the model, we can move the states to a separate location. All the models, before making a prediction, will fetch the states from that location. That separate location can be a master node in the distributed cluster or content store. Now the inference flow will look something like what's shown in *Figure 3.7*.

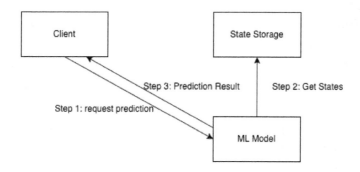

Figure 3.7 – Inference steps with states stored in a separate location

Once the model receives a prediction request, it will fetch the required model states from the state store. The states will then be applied to the model and predictions can be made from the model.

This approach has the following drawbacks:

- The additional calls to the state storage will add latency. We have to then consider the trade-off of updating the model states from the state storage periodically instead of for every client call.

- The size of the parameters can be big. So, we need to identify which parameters are safe to stay in the model without causing any side effects and decouple only the parameters that are most important and mutable.

This state storage will be common to all servers and the server can stay in sync by getting the parameters from the state storage from time to time.

If at a time instance, t, the state is updated, all the servers will be notified and they will take the new states from the state storage. We can avoid copying the model to the servers unless the model structure is changed and load the states only from the central store location.

To demonstrate this, let's take a very simple example using the decision tree we had before, and perform basic serving using the Flask API:

1. First of all, let's save the model and the model params separately using the following code:

```
from sklearn import tree
import pickle
X = [[0, 0, 2], [1, 1, 3], [3, 1, 3]]
Y = [0, 1, 2]
model = tree.DecisionTreeClassifier()
model.fit(X, Y)
with open("models/dt.pkl", "wb") as file:
    pickle.dump(model, file)
```

```
print(model.predict([[1, 1, 6]]))
params = model.__dict__
with open("state_store/dt_params.pkl", "wb") as param_
file:
    pickle.dump( params, param_file )
```

2. Next, we store the model in the `models` folder in a file called `dt.pkl`. We separately store the model parameters in the `state_store` folder in the `dt_params.pkl` file. We are training here with the same data as before:

```
X = [[0, 0, 2], [1, 1, 3], [3, 1, 3]]
Y = [0, 1, 2]
```

And we get the output of [1] from the line `print(model.predict([[1, 1, 6]]))`.

Now let's serve this model using the Flask API.

3. We will use the following code snippet to create serving endpoints using Flask. In this code sample, we create two REST API endpoints using `flask` for serving the model. One endpoint loads the model and parameters separately and the other endpoint loads the full model with parameters embedded as states of the model. We load the model using the `pickle` library and get the prediction of the data passed by the user. We explicitly indicate the type of the model after loading in the snippet `model: tree.DecisionTreeClassifier` as, by default, at the static compile time, the type is not resolved and we will not get autosuggestions for methods and attributes of the model during development. For predicting, we need to convert the JSON data to a numpy array using `json.loads(request.data)` as the `predict` method accepts numpy `ndarray` and returns numpy `ndarray` as a response. To encode numpy `ndarray` as JSON during the response, we create a custom encoder, NumpyEncoder. Without this, we will get an error as JSON, by default, does not know how to encode numpy `ndarray` as JSON:

```
import json
import pickle
from sklearn import tree
import numpy as np
from flask import Flask, jsonify, request
class NumpyEncoder(json.JSONEncoder):
    def default(self, obj):
        if isinstance(obj, np.ndarray):
            return obj.tolist()
        return json.JSONEncoder.default(self, obj)
app = Flask(__name__)
```

```
@app.route("/predict-with-params", methods=['POST'])
def predict_loading_params():
    with open("dt.pkl", "rb") as file:
        model: tree.DecisionTreeClassifier = pickle.
load(file)
        # Loading params from param store
        param_file = open("../state_store/dt_params.
pkl", "rb")
        params = pickle.load(param_file)
        print(params)
        model.__dict__ = params
        X = json.loads(request.data)
        print(X)
        response = model.predict(X)
        return json.dumps(response, cls=NumpyEncoder)
@app.route("/predict-with-full-model", methods=['POST'])
def predict_with_full_model():
    with open("dt.pkl", "rb") as file:
        model: tree.DecisionTreeClassifier = pickle.
load(file)
        X = json.loads(request.data)
        print(X)
        response = model.predict(X)
        return json.dumps(response, cls=NumpyEncoder)
app.run()
```

We load the model every time in the preceding dummy example. In reality, we might use a singleton service to return the model to avoid expensive I/O operations for each call. After running the application, we will see output like the following:

```
* Serving Flask app "flaskApi" (lazy loading)
* Environment: production
  WARNING: This is a development server. Do not use it
in a production deployment.
  Use a production WSGI server instead.
* Debug mode: off
* Running on http://127.0.0.1:5000/ (Press CTRL+C to
quit)
```

Please note the URL `http://127.0.0.1:5000/` as we will be using it in Postman. Postman is a REST API client.

We have two endpoints: `/predict-with-full-model`, which will do the prediction using the full model, and the other endpoint is `/predict-with-params`, which will not use the static model – it will fetch the parameters from the param store and fill the model with new parameters. We have moved the `dt.pkl` file to the serving folder. Now let's look at the following two cases:

- Serving does not load the params from the param store
- Serving loads the params from the param store

Serving without params from the param store

In this case, the application loads the full model from the `dt.pkl` file and makes a prediction. The drawback is every time the model updates, we need to replace the model in every server. In our example scenario, we are using only one server, but in reality, if we needed to serve the model for millions of users, we would need to use multiple servers:

1. Let's try to call the endpoint from Postman, as shown in *Figure 3.8*.

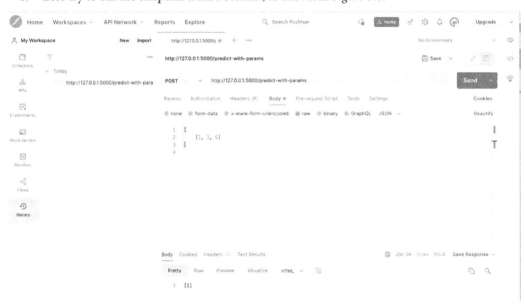

Figure 3.8 – Calling the endpoint predict-with-params from Postman

We see from *Figure 3.8*, that we got the response `[1]`. We will get the same result from both endpoints, as we copied the model from the first training to the model server and the params to the param store.

2. Now let's try to retrain the model with different data. The training code now looks like the following:

```
X = [[8, 4, 2], [5, 1, 3], [5, 1, 3]]
Y = [2, 1, 0]
model = tree.DecisionTreeClassifier()
model.fit(X, Y)
```

Now we get the response `[0]` from the line `print(model.predict([[1, 1, 6]]))`.

3. Now let's invoke both endpoints from Postman.

First, let's call the endpoint `"/predict-with-full-model"`. This model loads the full model along with the parameters from the model server. As we have not copied the updated model to the model server, we still get the output `[1]` for the input `[[1, 1, 6]]`, as shown in *Figure 3.9*.

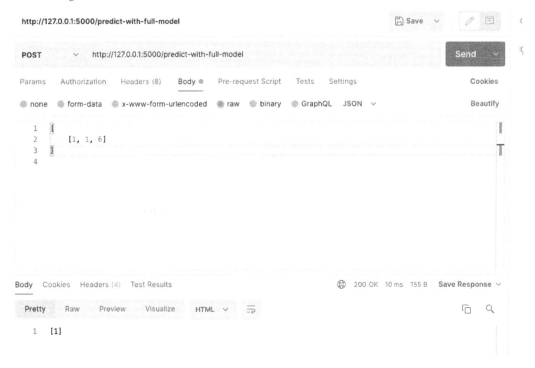

Figure 3.9 – Calling the endpoint "/predict-with-full-model" does not reflect the second training

4. Now let's call the other endpoint, which loads the params from the param store. The response is shown in *Figure 3.10*. Here, we get the response `[0]` from the served model. This is exactly the same as we got during testing after the second training was done.

Figure 3.10 – Calling the endpoint "/predict-with-params" reflects the updates due to the second training

We notice that the parameter decoupling helps to add consistency to the response.

Parameter decoupling can be more useful in neural networks. In neural networks, weights and biases are used as states and impact predictions. We can move these states to a state store.

5. For example, let's take the DNN model we saw earlier and save the model using the following code snippet:

```
model.save('saved_model')
```

This will save the model in a folder named saved_model, as shown in *Figure 3.11*.

Figure 3.11 – Saved Keras DNN model in the local workspace

We notice in *Figure 3.11* that the `saved_model` folder contains a file called `saved_model.pb`, which stores the structure of the model. The `variables` folder contains the weights and biases of the model.

6. When we want to load the model, we can use the following command:

```
model: tf.keras.models.Sequential = tf.keras.models.load_
model('saved_model')
```

It will load the model structure from the `saved_mode.pb` file and load the weights from the `variables` folder.

7. We can just save the weights separately, using the following command:

```
model.save_weights('mnist_weights')
```

This will save the weights and create two files in the same directory:

- mnist_weights.data*

- Mnitst_weights.index

8. If we want to replace the weights with weights from these two files, we need to use the following command:

```
model.load_weights("mnist_weights")
```

The whole code snippet for loading the model and weights is as follows:

```
import tensorflow as tf
import tensorflow_datasets as tfds
(ds_train, ds_test), ds_info = tfds.load(
    'mnist',
    split=['train', 'test'],
    shuffle_files=True,
    as_supervised=True,
    with_info=True,
)
def normalize_img(image, label):
  """Normalizes images: `uint8` -> `float32`."""
  return tf.cast(image, tf.float32) / 255., label
ds_train = ds_train.map(
    normalize_img, num_parallel_calls=tf.data.AUTOTUNE)
ds_train = ds_train.cache()
```

```
print(ds_info.splits['train'].num_examples)
ds_train = ds_train.shuffle(ds_info.splits['train'].num_
examples)
ds_train = ds_train.batch(128)
ds_train = ds_train.prefetch(tf.data.AUTOTUNE)
ds_test = ds_test.map(normalize_img)
ds_test = ds_test.batch(128)
model: tf.keras.models.Sequential = tf.keras.models.load_
model('saved_model')
model.summary()
preds = model.predict(ds_test)
print(preds[0]) # prediction without loading weights
separately
model.load_weights("mnist_weights")
preds = model.predict(ds_test)
print(preds[0]) # prediction with loading weights
separately
```

We see that there are two `print(pred[0])` statements. Before the last `print` statement, we load the weights of the model from the latest training. The outputs of the two `print` statements do not match. The output from the first `print` statement is shown here:

```
[2.1817669e-02 2.6139291e-04 6.7319870e-01 1.5672690e-01
2.3953454e-04
1.8283235e-02 2.0535434e-02 9.9055481e-04 1.0748719e-01
4.5936121e-04]
```

The output from the second `print` statement is different than the previous one:

```
[1.2448521e-03 4.2022881e-03 7.5482649e-01 2.8224552e-02
2.5956056e-04
5.9235198e-03 1.8221477e-01 4.1858567e-04 2.2648767e-02
3.6615722e-05]
```

These outputs may be different when you run your program. The point here is during the second prediction, we load the model weights from a separate decoupled location. This location could be a content server or web server. Before making predictions, the model will load the weights from this location and all the servers will stay in sync as this state store will be common to all the servers.

Summary

In this chapter, we have learned about stateful and stateless functions in more detail. We have seen how stateful serving can be problematic, causing impediments to scalable and resilient serving. Stateful serving can also violate fundamental computer science principles by causing servers to be out of sync. The response from servers will not be consistent once they are out of sync, violating the consistency principle.

We have discussed different kinds of states in machine learning models and how they impact inference. We also discussed that these models, with all these states, can be a big barrier to resilient and scalable serving.

We have seen some techniques to decouple states from ML models and have tried some demos using some dummy models by serving using the Flask API server.

In the next chapter, we will learn about continued model evaluation. We will discuss what the continuous model evaluation pattern is and why it is necessary, along with examples.

4

Continuous Model Evaluation

In this chapter, we will discuss continuous model evaluation, our second model serving pattern based on the model serving principle. Whenever we serve a model, we need to keep in mind that this model will underperform after a certain period. As data is growing continuously and new data patterns appear regularly, a model that has been trained on old data will very soon become stale. In this chapter, we will introduce you to continuous model evaluation, why it is needed, the techniques of continuous model evaluation, and example scenarios. At a high level, we will discuss the following topics:

- Introducing continuous model evaluation
- The necessity of continuous evaluation
- Continuous model evaluation use cases
- Evaluating a model continuously
- Monitoring model performance when predicting rare classes

Technical requirements

In this chapter, you will need to have access to a Python IDE, the code from *Chapter 3*, and the code from *Chapter 4*. All this code can be found in this book's GitHub repository in the respective folders. The link to this chapter's code is `https://github.com/PacktPublishing/Machine-Learning-Model-Serving-Patterns-and-Best-Practices/tree/main/Chapter%204`.

You do not have to install any additional libraries for this chapter. However, if you have difficulty accessing a library, make sure you install the library using the `pip3` command. This chapter will require the `scikit-learn` library and `flask`. Please install `scikit-learn` using `pip3 install scikit-learn` and install `flask` using `pip3 install Flask`.

Introducing continuous model evaluation

In this section, you will be introduced to continuous model evaluation.

The continuous evaluation of a model is a strategy in which the model's performance is evaluated continuously to determine when the model needs retraining. This is also known as monitoring. Usually, there is a dashboard to monitor the values of different metrics over time. If the value of a metric falls below a certain threshold, indicating a performance drop, then there is usually a mechanism to notify the users.

In software engineering, after an application has been deployed, there are some service/operational metrics and business metrics that are monitored continuously to understand whether the application is performing correctly and meeting the business requirements.

For example, let's imagine that we have launched a web application for users to read books online. Let's think about what metrics need to be monitored to determine whether this application is operating well and achieving business success.

Some of the service/operational metrics could be the following:

- *Latency*: This will determine how much time the users need to wait to get a response after a request has been made. This metric is critical for target customer satisfaction.

- *Error rate*: This is the count of client-side/server-side errors over a time frame that is being used for evaluation. This metric helps to signal to developers that a malfunction is going on. Then, the developer can start debugging to understand what kind of bugs are occurring and what kind of improvements can be made to make the application more user-friendly and error-free.

- *Availability*: Availability is measured as a percentage. We set a time frame and measure what percentage of time the application was down over that last time frame. Then, we subtract that downtime percentage from 100%. An availability of nearly 100% is desirable.

Some of the business metrics could be the following:

- *Registration count*: This is the count of the number of new members registered to the application over the last time frame.

- *Account Deletion Count*: This specifies how many users deleted their accounts over the last time frame.

- *Average Time Spent*: This is the average time spent daily by users of the application.

Service metrics help to monitor the health of the application, while business metrics help to monitor the business impact of the application. In software engineering, continuous evaluation is considered a key element to ensure the success of software or an application.

These metrics are usually visualized on a dashboard for quick lookup. If the metrics are under or over the acceptable limit, then, usually, a notification is sent to the developers. For some metrics, the value can be bad if it goes below the threshold, while for other metrics, the value can be bad if it goes above the threshold:

- *A metric value lower than the threshold is bad*: For these metrics, the higher the value is, the better; for example, with throughput and availability, we want *high* throughput and *high* availability. Therefore, if these values drop below a predefined threshold, we need to be concerned.

- *A metric value higher than the threshold is bad*: For these metrics, the lower the value is, the better; for example, with latency and error count, we want *low* latency and *few* errors. Therefore, if these values go above a certain threshold, we need to take action.

Most big corporate companies assign a significant amount of developers' time to monitor the service their team is managing. You can learn more about the monitoring support provided by Google Cloud at `https://cloud.google.com/blog/products/operations/in-depth-explanation-of-operational-metrics-at-google-cloud` and Amazon AWS monitoring dashboard support, monitoring metrics, and alarms at `https://docs.aws.amazon.com/AmazonCloudWatch/latest/monitoring/CloudWatch_Dashboards.html`.

Let's think about how the metrics and continuous evaluation of the application using these metrics help the success of software:

- *Operational perspectives*: If an operational metric exceeds the threshold constraint, then we know our application needs to be fixed and redeployed. The developers can start digging deep to localize bugs and optimize the program.

- *Business perspectives*: By monitoring business metrics, we can understand the business impact and make the necessary business plan to enhance business impact. For example, if the registration count is low, then the business could think of a marketing strategy and developers could think about making registration more user-friendly.

So, we now know that monitoring can help us. But what should we look out for?

What to monitor in model evaluation

We saw some of the metrics that can be observed in software applications in the previous section. Now, the question is, to continuously evaluate a model, what metrics should we monitor? In model serving, the operational metrics can remain the same; as we discussed in *Chapter 1, Introducing Model Serving*, model serving is more of a software engineering problem than a machine learning one. We are not particularly interested in discussing those metrics. However, we will discuss some operational metrics that are related to the model, such as bad input or the wrong input format.

For continuous model evaluation, the most critical goal is to monitor the performance of the model in terms of inference. We need to know whether the inferences made by the model have more and more prediction errors. For example, if a model doing binary classification of cats and dogs starts failing to classify most of the input data, then the model is of no use anymore.

To evaluate model performance, we should use the same prediction performance metric that we used during training. For example, if we used accuracy as a metric while training the model, then we should also use accuracy as a metric to monitor during the continuous evaluation of the model after deployment. If we use a different metric, that can work but will create confusion.

For example, let's say during training we trained the model and stopped training after the accuracy was 90%. Now, if we use a different metric, such as the F1 score, during continuous evaluation, we face the following problems:

- What is the ideal threshold value to set for the F1 score for the continuous evaluation of this model?

- How are accuracy and the F1 score related to the signal for when to start retraining?

To get rid of this confusion, we should use the same metric during training and continuous evaluation. However, we can monitor multiple parameters during continuous evaluation, but to make a retraining decision, we should use the value of the parameter we used during training.

Challenges of continuous model evaluation

In software applications, monitoring metrics can be configured without human intervention. Both the computation of operational metrics and business metrics can be calculated from the served application itself. For example, to monitor latency, we can start a timer once an application is called by the user and measure the time taken to get the response. This way, from the application itself, we can get the necessary data to compute latency. For business metrics such as counting registrations, we can increment a counter whenever a new registration succeeds.

However, for model evaluation, the ground truth for generating metrics needs to be collected externally.

For example, let's say the metric is **Square Error** (**SE**). To measure this metric, the formula is as follows:

$$\sum_{i=1}^{n} (y_i - y_i')^2$$

Here, y_i is the actual value or ground truth for an input instance and $y'i$ is the predicted value. Now, let's say we have a model for predicting house prices. The model is predicting the features of a house and it is y'_i. Now, how can we know the actual price? Sometimes, the ground truth may not even be available at the time of prediction. For example, in this case, to know the actual price, y_i, we have to wait until the house is sold. Sometimes, the actual value may be available, but we need human intervention to collect those ground truth values.

Sometimes, knowing the actual value may not be legal. For example, let's say a model predicts some personal information about users – accessing that private information might be disallowed in some cultures. Let's say some models determine whether a patient has a particular disease, whether a person has particular limitations, and so on. If we access that information without legal processes, we might be violating laws.

Sometimes, the process of collecting the ground truth may be biased. For example, let's say we have a model that predicts the rating of a particular product. To collect the ground truth, we might ask some human labelers to rate those. This can be prone to high bias, making the continuous evaluation biased as well.

Therefore, the problems with getting the ground truth to generate the model are as follows:

- The ground truth may not be available at the time of prediction
- Human intervention is needed and is prone to error sometimes
- Sometimes, knowing the actual value may not be legal
- There is the possibility of bias

In this section, we have discussed what continuous model evaluation is and what we monitor during continuous evaluation, along with an intuitive explanation of some of the metrics that need monitoring. In the next section, we will discuss the necessity of continuous evaluation more elaborately.

The necessity of continuous model evaluation

In the previous section, you were introduced to continuous model evaluation, challenges in continuous model evaluation, and an intuitive explanation of some metrics that can be used during continuous evaluation. In this section, you will learn about the necessity of continuous model evaluation. Once a model has been deployed after successful training and evaluation, we can't be sure that the model will perform the same continuously. The model performance will likely deteriorate over time. We also need to monitor the model for 4XX (client-side errors) and 5XX errors (server-side errors). Monitoring these error metrics and doing necessary fixes are essential for maintaining the functional requirements of serving:

- Monitoring errors
- Deciding on retraining
- Enhancing serving resources
- Understanding the business impact

Let's look at each in turn.

Monitoring errors

Monitoring errors is essential to providing smooth model serving for users. If users keep getting errors, then they will face difficulties in getting the desired service from the model.

Errors can mainly be classified into two groups: *4XX* and *5XX* errors. We will discuss these errors briefly in the following subsections.

4XX errors

All kinds of client-side errors fall into this category. Here, XX will be replaced by two digits to represent an actual error code. For example, the *404 (Not Found)* error code indicates the serving endpoint you have provided is not available to the outside world. So, if you see this error, then you need to start debugging why the application is not available. You need to try to access the endpoint outside your organization's VPN to ensure the endpoint is available. *401 (Unauthorized)* indicates that the client is not authenticated to use the serving endpoint. If you see this error, then you need to start debugging whether the endpoint has some bad authentication applied and whether the API can authenticate the user correctly. *400 (Bad request)* indicates that the client is not calling the serving endpoint properly. For example, if the payload is passed in the wrong format while calling the inference API, you might get a 400 error. This kind of error will be more common in model serving as the model will expect the input features in a particular format.

To understand 400 errors, let's take the serving code from *Chapter 3*, *Stateless Model Serving*, and modify the "/predict-with-full-model" endpoint, as follows:

1. Let's add an `if` block to abort the program and raise a 400 error for the wrong dimension:

    ```
    from flask import abort
    @app.rout"("/predict-with-full-mod"l", methods'['PO'T'])
    def predict_with_full_model():
        with ope"("dt.p"l"" ""b") as file:
                model: tree.DecisionTreeClassifier = pickle.
    load(file)
                X = json.loads(request.data)
                print(X)
                if len(X) == 0 or len(X[0]) != 3:
                        abort(400" "The request feature
    dimension does not match the expected dimensi"n")
                response = model.predict(X)
                return json.dumps(response, cls=NumpyEncoder)
    ```

    We have imported `abort` to throw the errors. We have also added a very basic case of throwing a 400 error when the feature dimension in the input does not match the feature dimension used during training.

2. Next, we must test whether the error is occurring because we're passing data of the wrong dimension during the call of the prediction endpoint.

    Recall that we used the following data to train a basic decision tree model to demonstrate serving in *Chapter 3*:

```
X = [[0, 0, 2], [1, 1, 3], [3, 1, 3]]
Y = [0, 1, 2]
```

Notice that the input feature is a list of three training instances. Each training instance has a length of 3. After training the model, we need to pass the list of features following the same dimension criteria. This means that each input we pass for training must be a length of 3 as well. As model deployment engineers, we need to throw proper errors in these cases.

3. Now, let's add the following lines to throw a 400 error if these dimension criteria are not matched:

```
if len(X) == 0 or len(X[0]) != 3:
            abort(400" "The request feature
    dimension does not match the expected dimensi"n")
```

The logic to throw here is very basic. We are only checking the first instance for size three in `len(X[0] != 3`. However, you need to check the size of all the instances. Usually, machine learning engineers that have less access to software engineering neglect these errors being thrown and during serving, proper health monitoring becomes a tedious job.

4. Now, we will play with the output in Postman, as in *Chapter 3*. This time, we will pass an input of a bad dimension:

Figure 4.1 – A 400 error is thrown for passing a request with an input of a wrong dimension

In *Figure 4.1*, we passed an input of `[[1, 1]]`. Here, the length of the first feature is two but the model requires it to be three. So, we get a 400 error. The response that we get is as follows:

```
<!DOCTYPE HTML PUBLI" "-//W3C//DTD HTML 3.2 Final//"N">
```

```
<title>400 Bad Request</title>
<h1>Bad Request</h1>
<p>The request feature dimension does not match the
expected dimension</p>
```

Here, the title (`<title></title>`) represents the error code and the paragraph (`<p></p>`) represents the message that we have passed. We should provide a clear message to the client. For example, in the error message, we could also provide the dimension information to indicate what was expected and what was passed. This would give a clear suggestion to the user on how to fix it.

5.   For example, if we modify the code in the following way to add a more verbose error message, it will give clear instructions to the client:

```
if len(X) == 0 or len(X[0]) != 3:
            message ="""""
The request feature dimension does not match the expected
dimension.
The input length of {X[0]} is {len(X[0])} but the model
expects the length to be 3."""""
            abort(400, message)
```

This time, after making a request with the same input as before, we get the following error message:

```
<!DOCTYPE HTML PUBLI" "-//W3C//DTD HTML 3.2 Final//"N">
<title>400 Bad Request</title>
<h1>Bad Request</h1>
<p><br>The request feature dimension does not match the
expected dimension.<br>The input length of [1, 1] is 2
but the model expects the length to be 3.
<br></p>
```

Now, the error code and message will help the developer improve the functionality of the application. How can monitoring errors help developers? Developers can enhance the API design and make plans in the following ways:

•   Use a modeling language for API calls to build a request and ensure all the fields in the dimension are required. For example, using **smithy** (`https://awslabs.github.io/smithy/`), you can specify the input structure in an API call. However, it will still be in the client code. The code will not be able to hit the model serving code and the error can still come from the client code calling the APIs. For example, to specify the input list size as 3, you can use the following code snippet in `smithy` to define the model:

```
@length(min: 3, max: 3)
```

```
list X {
      member: int
}
```

Here, we created an input for the API, X, which is a list of integer numbers; the list's size needs to be 3:

- Provide a user interface with the required fields for the users. In this way, we can add a sanity check to the client interface to ensure valid data is passed with the request.

- Create a script to automate the process of collecting input data for inference and formatting and then sending requests.

To learn more about REST API design and best practices, please visit https://masteringbackend. com/posts/api-design-best-practices. All these ideas can come onto the discussion table to enhance serving through monitoring these metrics. To learn more about 4XX errors, please follow this link: https://www.moesif.com/blog/technical/monitoring/10-Error-Status-Codes-When-Building-APIs-For-The-First-Time-And-How-To-Fix-Them/.

5XX errors

5XX errors occur when there is some error from the server side. Let's look at one of the most common errors: *500 (Internal Server Error) error*.

If there is something wrong during processing in the server-side code, then this error will appear. If there is an exception in the server-side code, it will throw a 500 error. To demonstrate this, let's call the previous API endpoint, "/predict-with-full-model", with the following data:

```
[
["a", "b", "c"]
]
```

Here, we passed some strings, ["a", "b", "c"], instead of the numbers. The dimension of this input is 3, which is why this does not enter the conditional block:

```
if len(X) == 0 or len(X[0]) != 3:
            message = f"""
The request feature dimension does not match the expected
dimension.
The input length of {X[0]} is {len(X[0])} but the model expects
the length to be 3.
"""

            abort(400, message)
```

That's why we won't see the 400 error. However, this time, the input list is not a list of numbers. If we pass this input to Postman, we will see the response of the 500 error, as shown in *Figure 4.2*:

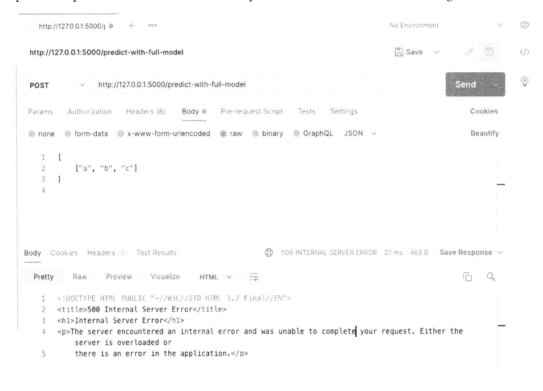

Figure 4.2 – An 500 error is thrown if the server code fails to process the client request

Now, the developer can go to the server and look at the log that is causing this error:

```
array = np.asarray(array, order=order, dtype=dtype)
  File "/Library/Frameworks/Python.framework/Versions/3.8/lib/
python3.8/site-packages/numpy/core/_asarray.py", line 83, in
asarray
    return array(a, dtype, copy=False, order=order)
ValueError: could not convert string to float: 'a'
```

We get the preceding log in the terminal in the PyCharm IDE. Now, the developer knows why this error is appearing from the "ValueError: could not convert string to float: 'a'" error message. The client is passing some strings that could not be converted into floating-point numbers by the server code. What actions can the developers take? The developers can add a sanity check to the input to throw a **400 (Bad Request)** error to the client if the type of the input data does not match the expected type. This **500 (Internal Server Error)** is more common during model serving. There are a few other 5XX errors, such as 502 indicating issues in the gateway serving the API, 503 indicating unavailability of service, and 504 indicating timing out from the service. All these errors will help the developer to monitor server health.

Deciding on retraining

The crucial use case of continuous mode evaluation is to identify when the model is underperforming and when we should go for retraining. The goal of continuous model evaluation is different from that of software engineering application serving. When we train a model, we optimize the model to use a metric to understand whether the trained model is strong enough to make predictions. Then, we deploy the model to the server. However, the model will start underperforming as time goes on. A few reasons for it underperforming are as follows:

- *The volume of unseen data with different distributions becomes high*: By unseen data, I mean the data that was *not* used while training the model and has changed in the distribution of the data. In machine learning, this problem is widely known as **data drift**, as described at https:// arize.com/model-drift/. As time goes on, the volume of unseen data will increase. We know that, during training, we use a major share of available data for training for some rules of thumb, such as the 80-20 split rule. Therefore, with this new volume of unseen data, the past model will behave like an underfitted model. One of the reasons for underfitting is not using sufficient data during training.

- *The arrival of new patterns of data*: Let's say we are using a classification model to recognize the names of fruits in a garden. Let's say the garden has only five fruits. So, the model is trained to classify these five fruits. Now, from the beginning of this year, the garden owner started to grow a sixth type of fruit. So, the old model becomes unusable now, as it will always misclassify the sixth fruit as one of the other five classes. We will be able to catch this degradation in performance through continuous monitoring. One thing we need to be careful of is that if the model fails to classify a rare class, then we might not notice a significant drop in performance. Therefore, we have to take care of rare classes during monitoring. One strategy could be to add more of a penalty by adding more weight during the misclassification of rare classes during evaluation.

- *A change of market condition*: A change of market condition may make a lot of features and their corresponding target values misaligned. For example, at the beginning of the pandemic, stock prices suddenly dropped. So, if we predicted the stock price using the same model before the pandemic, we would get a huge error, and that would tell us to retrain the model. Similarly, during inflation, market prices experience turmoil. If we use the same model as before, we will get more errors. The ground truths have now changed; this problem is widely known as **concept drift** in machine learning, as described at https://arize.com/model-drift/. If we do not continuously monitor the model, then we will miss the opportunity to update the model and get wrong predictions.

- *A change of data calibration*: A change of data calibration indicates that the business decided on a different label for the same input feature. This is widely known as **prediction drift** in machine learning, as described at `https://arize.com/model-drift/`. For example, let's say we have a model to determine whether we should call a candidate for an interview for a particular position or not. This is a binary classification problem. This means that the label is either YES or NO. Let's say a subset of the data in 2021 and 2022 for the company is shown in the table in *Figure 4.3*. For the same features, the company invited candidates for interviews in 2021, but in 2022, the company decided to no longer invite candidates for the same position with the same features for interviews in 2022. So, the model that automatically makes this prediction will not work in 2022, even though the model was very strong in 2021:

Years of Experience (Years)	Major	Highest Degree	Decision 2021	Decision 2022
1	CS	MS	YES	NO
2	CS	BSc	YES	NO
1.5	CS	BSc	YES	NO
3	CS	BSc	YES	NO

Figure 4.3 – Interview call decisions based on the same features in 2021 and 2022

For the cases described here, we have to retrain the model. To make this decision, it helps automatically monitor the model metric continuously after deployment.

Enhancing serving resources

One of the usages of continuous model evaluation is to decide whether our serving resources need scaling. We need to monitor operational metrics such as availability and throughput for this. We can also monitor different errors such as **request timeout** errors, **too many requests** errors, and so on to determine whether our existing computation resource is no longer sufficient. If we get request timeout errors, then we need to enhance the computation power of the server. We might think about parallelizing computation, identifying bottlenecks, doing load testing, or using high-power GPUs or other high-power processors. If there are too many request errors, we need to horizontally scale the serving infrastructure and set a load balancer to drive requests to different servers to maintain an equal load on the servers.

To understand the load on our server, we can monitor the traffic count at different times of the day. Based on that, we can only enable high computation during peak times instead of keeping the resources available all the time.

Understanding business impact

The business success of a model depends on the business impact of the model. By continuously evaluating the model by monitoring the business metrics, we can understand the business impact of the model. For example, we can monitor the amount of traffic using the model daily to understand whether the model is reaching more clients. Based on the model's business goal, we can define the metrics and keep monitoring those metrics via a dashboard.

For example, let's assume we have a model to predict the estimated delivery time of food from a food delivery service provider app. The app needs to provide an estimated delivery time of food to the client when the client orders food using this app. Now, let's say the app developers want to understand the business impact of the model. Some of the business metrics they could monitor are as follows:

- *The number of users*: How many new users have started to use the app can be a good metric to understand whether the app has been able to reach more people.

- *Churn rate*: How many users have stopped using the app can indicate whether the prediction made by the model does not reflect the actual delivery time.

- *Average rating*: The average star rating of the app can also indicate whether customers are satisfied with the performance of the app.

- *The sentiment of reviews*: Customers can write reviews about the delivery process, such as whether the food arrived timely or not. This sentiment analysis of reviews can show whether customers are happy with the delivery time predictions of the model or not.

We can also monitor model metrics such as the MSE of predicted delivery times and actual delivery times and do correlation analysis with customer ratings, churn rate, new user signup rate, and other business metrics to understand the correlation of model performance with the business metrics.

In this section, we have seen the necessity of continuous model evaluation. We saw that continuous model evaluation is needed to keep an eye on server health, to understand the business impact, and, most importantly, to understand when a model starts underperforming.

Next, let's look at some of the most commonly used metrics so that we can use them wisely in continuous evaluation.

Common metrics for training and monitoring

In this section, we will introduce some common metrics for training and monitoring during continuous model evaluation. A detailed list of metrics used in training different kinds of machine learning models can be found here: `https://scikit-learn.org/stable/modules/model_evaluation.html`.

We shall discuss a few of them here to get a background understanding of the metrics.

Accuracy

Accuracy measures the percentage match of the predicted value and the actual value in a multilabel classification or multi-class problem. For example, if 70 predictions were made correctly out of 100 total predictions, then the accuracy would be 70%. Let's look at the following code snippet from scikit-learn for computing accuracy:

```
from sklearn.metrics import accuracy_score
y_pred = [0, 2, 1, 3]
y_true = [0, 1, 2, 3]
acc1 = accuracy_score(y_true, y_pred)
print(acc1) # Prints 0.5
acc2 = accuracy_score(y_true, y_pred, normalize=False)
print(acc2) # Prints 2
```

Here, acc1 is the accuracy as a percentage – that is, the total correct predictions divided by the total number of predictions made by the model. On the other hand, acc2 gives the actual number of correct predictions made. If we are not careful, the decision made by acc2 could be biased. For example, let's say the acc2 value is 70. Now, if the total number of predictions made is also 70, then we have 100% success. On the other hand, if the total number of predictions made is 1,000, then we have only 7% success.

In the preceding code, the value of acc1 is 0.5 or 50% because, in the y_true and y_pred arrays, two samples out of the four match. In other words, two out of four predictions were correct, and the value of acc2 is 2, indicating the concrete number of correct predictions.

Now, let's implement the same metric function outside the library using simple Python logic to understand the metric well. We are avoiding handling all the corner cases to highlight just the logic:

```
def accuracy_score(y_true, y_pred, normalize = True):
    score = 0
    size = len(y_pred)
    for i in range(0, size):
        if y_pred[i] == y_true[i]:
            score = score + 1
    if normalize:
        return score/size
    else:
        return score


y_pred = [0, 2, 1, 3]
y_true = [0, 1, 2, 3]
acc1 = accuracy_score(y_true, y_pred)
print(acc1) # Prints 0.5
```

```
acc2 = accuracy_score(y_true, y_pred, normalize=False)
print(acc2) # Prints 2
```

We get the same output as before from the preceding code snippet. It helps to understand how the metric is implemented and to understand the metric with more intuition. This is a simple metric that can be used to train a multi-label classification model.

Precision

Precision measures the ratio of true positives to the total number of positives predicted by the model. Some of the predicted positives are false positives. The formula for precision is $tp/(tp + fp)$. Here, tp refers to the true positive count and fp refers to the false positive count. From the formula, we can see that if fp is high, then the denominator will be high, making the value of ratio or precision low. So, intuitively, this ratio discourages false positives and optimizes precision indicates, reducing the false alarms from the model. Positive and negative indicates two labels in binary classification. In a multi-class classification problem, this is a little difficult. In that case, we have to take the average of considering one class positive and all other classes negative at a time. The lowest value of precision is 0 and the highest value is 1.0.

To understand precision better, let's look at the following code snippet from scikit-learn to compute the precision:

```
from sklearn import metrics
y_pred = [0, 1, 0, 0]
y_true = [0, 1, 1, 1]
precision = metrics.precision_score(y_true, y_pred)
print(precision) # Prints 1.0
```

From the preceding code snippet, we can see that the precision score that was computed using the scikit-learn API is 1.0 or 100%. The API considers 1 as the positive value and 0 as the negative value. So, if we manually compute the number of true positives and false positives, the total tp is 1 and fp is 0 because the only predicted positive is at index 1, which is also the true value. So, in this example, we only have a true positive, and there are no false positives in the response. Therefore, the precision is 1 or 100%.

To understand this logic better, let's implement it in Python in the following code snippet. We are excluding all the corner cases just to show the logic:

```
def precision(y_true, y_pred):
    tp = 0
    fp = 0
    size = len(y_true)
```

```
        Positive = 1
        Negative = 0
        for i in range(0, size):
                if y_true[i] == Positive:
                        if y_pred[i] == Positive:
                                tp = tp + 1
                else:
                        if y_pred[i] == Positive:
                                fp = fp + 1

        return tp/ (tp + fp)

y_pred = [0, 1, 0, 0]
y_true = [0, 1, 1, 1]
precision = precision(y_true, y_pred)
print(precision) # Print 1.0
```

In the preceding code snippet, we implemented the basic logic of computing precision and we got the same result for the same input that we received from scikit-learn.

Here, we got a precision score of 1.0. We might be biased to think that we have a very perfect model. However, if we look at the y_true and y_pred arrays, we will notice that only 50% of the y_true values were predicted correctly. So, there is a risk in using precision alone as the scoring metric. If the model has a lot of false negatives, that means the model cannot recognize the negative cases at all, so the model will be very bad, even though the precision will be high. Consider a model that just has a single line of code, return 1, where there is no logic inside it. The model will have a precision of 100%, which can give us a biased interpretation of the model.

Recall

Recall is a metric that gives the ratio of true positive counts to the sum of the true positive counts and false negative counts. The formula of recall is tp/ (tp + fn). Here, tp stands for true positives and fn stands for false negatives. False negatives refer to predictions where the true value is 1 but the prediction is 0 or negative. As these predictions are false or wrong, they are known as false negatives.

If the value of fn is high, then the value of the denominator in the formula for precision will also be high. Therefore, the ratio or recall will be lower. In this way, recall discourages false negatives and helps to train a model that will be optimized to produce fewer false negatives.

To understand the recall metric better, first, let's see the following `scikit-learn` code snippet computing the recall:

```
from sklearn import metrics
y_pred = [0, 1, 0, 0]
y_true = [0, 1, 1, 1]
recall = metrics.recall_score(y_true, y_pred)
print(recall) # Prints 0.3333333333333333
```

Now, the recall score is ~0.33 for the same data shown in the example that computed the precision. The precision failed to tell us the model that's predicting these values is a bad model. However, from the recall, we can tell that the model that was predicting a lot of false negatives is bad.

Now, to understand how the recall metric is generated, let's see the implementation by using raw Python code, as follows:

```
def recall(y_true, y_pred):
    tp = 0
    fn = 0
    size = len(y_true)
    Positive = 1
    Negative = 0
    for i in range(0, size):
        if y_true[i] == Positive:
            if y_pred[i] == Positive:
                tp = tp + 1
            else:
                fn = fn + 1

    return tp/ (tp + fn)

y_pred = [0, 1, 0, 0]
y_true = [0, 1, 1, 1]
recall = recall(y_true, y_pred)
print(recall)
```

Here, we can see the same output as before. Notice that if the `y_pred` value is negative, then we do not do anything. This gives us a clue that there is some issue with the recall metric as well.

To understand this, let's run the following `scikit-learn` model with the following input:

```
y_pred = [1, 1, 1, 1, 1]
y_true = [0, 0, 0, 0, 1]
recall = metrics.recall_score(y_true, y_pred)
print(recall)
```

We get a recall of 100% from the preceding code snippet. However, notice that out of five instances, the model only predicted one instance correctly, indicating a success of 20%. Out of the five predictions, only one was a true positive and the other four were false positives. So, we can see that the recall model gives a very biased interpretation of the model's strength if the model makes a lot of false positives.

F1 score

From the discussions on precision and recall, we have seen that precision can optimize a model to be strong against false positives and recall can optimize the model to be strong against false negatives. **F1 score** combines these two goals to make a model that is strong against both false positives and false negatives. The F1 score is the harmonic mean of precision and recall. The formula for the F1 score is `2/(1/Precision + 1/Recall)`. There are some variants of the F1 score based on weights applied to precision. The actual formula is `(1 + b^2)/ (b^2/recall + 1/precision)` `=> (1 + b^2)*precision*recall/(b^2*precision + recall)`. This beta parameter is used to differentiate the weight given to precision and recall. If beta > 1, then precision has a lower weight as the weight will cause precision to contribute more to the denominator in the preceding formula, making the share of precision more than its actual value. If beta < 1, then recall has more weight compared to precision as precision will contribute less than its actual value. If beta = 1, then both precision and recall will have the same weights.

Now, to understand how the F1 score utilizes the best aspects of precision and recall, let's compute the F1 score of the preceding data using scikit-learn:

```
from sklearn import metrics
y_pred1 = [0, 1, 0, 0]
y_true1= [0, 1, 1, 1]

F1 = metrics.f1_score(y_true1, y_pred1)
print(F1) # Prints 0.5
```

```
y_pred2 = [1, 1, 1, 1, 1]
y_true2 = [0, 0, 0, 0, 1]

F1 = metrics.f1_score(y_true2, y_pred2)
print(F1) # Prints 0.33333333333333337
```

Notice that for both sets of data, the **F1 score** now better represents the actual prediction power of the models. For the first dataset, for y_pred1 and y_true1, the **F1 score** is 50%. However, the precision was 100%, even though the prediction success was 50%.

For the second dataset, the value of the F1 score is ~0.3333. However, the recall was 100%, even though the prediction success was only 20%. Therefore, we can see that the F1 score is less biased in representing the power of the model. So, while monitoring for continuous model evaluation, we need to be careful when using mere precision or recall. We might be fooled into assuming the model is doing well, but it is not. It's safer to use the F1 score in place of them.

There are many other metrics for classifications. We encourage you to go through all these metrics and try to implement them to understand which metrics are good for your business case. To understand model evaluation, understanding these metrics from the basics will help you to monitor and evaluate wisely.

In the next section, we will see some use cases of continuous model evaluation. We will demonstrate through examples how continuous model evaluation should be an integral part of models serving different business goals.

Continuous model evaluation use cases

In this section, we will see some cases to demonstrate when model performance monitoring is essential.

To understand when continuous model evaluation is needed, let's take an example case of a regression model to predict house prices:

1. First, let's take dummy example data showing house prices from January 2021 to July 2021.

    When we deploy the regression model, we have to continuously monitor regression metrics such as **Mean Square Error (MSA)** and **Mean Absolute Percentage Error (MAPE)** to understand whether our model's prediction started deviating a lot from the actual value. For example, let's say that a regression model to predict house prices was trained in January using the data shown in the table in *Figure 4.4*:

#Bedroom	#Bathroom	Size	Lot Size (Acre)	Price in Jan 2021	Price in February 2021	Price in March 2021	Price in April 2022	Price in May 2021	Price in June 2021	Price in July 2021
3	3	2000	.2	350k	360k	370k	380k	390k	410k	420k
4	3	2200	.3	430k	445k	460k	475k	500k	515k	530k
5	4	3000	.4	550k	570k	590k	610k	630k	650k	670k
2	2	1800	.2	300k	305k	310k	315k	325k	330k	340k

Figure 4.4 – Mock house price data against features in Jan 2021 and July 2021

In the table in *Figure 4.4*, we have also shown the hypothetical price change of houses until July. Let's assume the model was developed with the data until January and was able to make a perfect prediction with zero errors in January. Though these predictions were perfect for the market in January, these were not good predictions from February to July.

2. Next, we must compute the MSE of the predictions of the house prices in different months by the model and plot.

We can plot the MSE using the following code snippet:

```
from sklearn.metrics import mean_squared_error
import matplotlib.pyplot as plt
import numpy as np

predictions = [350, 430, 550, 300]
actual_jan = [350, 430, 550, 300]
actual_feb = [360, 445, 570, 305]
actual_mar = [370, 460, 590, 310]
actual_apr = [380, 475, 610, 315]
actual_may = [390, 500, 630, 325]
actual_june = [410, 515, 650, 330]
actual_july = [430, 530, 670, 340]

mse_jan = mean_squared_error(actual_jan, predictions)
mse_feb = mean_squared_error(actual_feb, predictions)
mse_mar = mean_squared_error(actual_mar, predictions)
```

```
mse_apr = mean_squared_error(actual_apr, predictions)
mse_may = mean_squared_error(actual_may, predictions)
mse_june = mean_squared_error(actual_june, predictions)
mse_july = mean_squared_error(actual_july, predictions)

errors = np.array([mse_jan, mse_feb, mse_mar, mse_apr,
mse_may, mse_june, mse_july])
plt.plot(errors)
plt.ylabel("MSE")
plt.xlabel("Index of Month")
plt.show()
```

We will see a curve like the one shown in *Figure 4.5*:

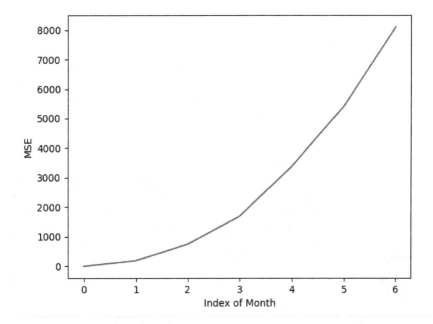

Figure 4.5 – Change of MSE from a hypothetical model until July 2021 that was trained in January 2021

From *Figure 4.5*, we can see that the MSE of prediction is increasing. We can show this monitoring chart on a dashboard to decide on retraining. We can see from the chart that the model performance is deteriorating as we didn't update the model. The model is no longer usable and is dangerous to use after a certain time.

Now, let's consider the case of a classification model. Let's say it makes a binary decision on whether a person is eligible for a loan or not:

1. First, let's take some hypothetical training data. Let's say the feature to train the model in January 2021 was as shown in the table in *Figure 4.6*:

Monthly Income	Debt	Can Get Loan
10k	30k	Yes
15k	30k	Yes
20k	35k	Yes
25k	40k	Yes
30k	75k	Yes
21k	40k	Yes
30k	45k	Yes
40k	100k	Yes
17k	12k	Yes
19k	20k	Yes

Figure 4.6 – Hypothetical feature set used in January 2021 to decide whether a person can get a loan

For simplicity, let's assume all 10 people were worthy of getting a loan in January 2021.

2. Now, let's assume these 10 people stay with the same financial conditions until July, but due to inflation and market conditions, the bank has slowly tightened the rules on giving loans over the months until July. Let's assume that a person's worthiness of getting a loan over the next 6 months looks like the table in *Figure 4.7*:

Feb	Mar	Apr	May	Jun	Jul
No	No	No	No	No	No
Yes	No	No	No	No	No
Yes	Yes	Yes	No	No	No
Yes	Yes	Yes	Yes	No	No
Yes	Yes	Yes	Yes	No	No
Yes	Yes	Yes	Yes	Yes	No
Yes	Yes	Yes	No	No	No
Yes	Yes	Yes	Yes	Yes	No
Yes	Yes	No	No	No	No
Yes	Yes	No	No	No	No

Figure 4.7 – Change of loan worthiness until July 2021

Notice that the actual loan worthiness of every person was lost by July 2021. However, as the model was trained in January 2021, and was passed the same features, the model will still predict that all 10 people are worthy of getting a loan, making the model unusable and unreliable.

3. Let's show the accuracy of the model over the months using the following code snippet:

```
from sklearn.metrics import accuracy_score
import matplotlib.pyplot as plt
import numpy as np

predictions = [1, 1, 1, 1, 1, 1, 1, 1, 1, 1]
actual_jan = [1, 1, 1, 1, 1, 1, 1, 1, 1, 1]
actual_feb = [0, 1, 1, 1, 1, 1, 1, 1, 1, 1]
actual_mar = [0, 0, 1, 1, 1, 1, 1, 1, 1, 1]
actual_apr = [0, 0, 1, 1, 1, 1, 1, 1, 0, 0]
actual_may = [0, 0, 0, 1, 1, 1, 0, 1, 0, 0]
actual_jun = [0, 0, 0, 0, 0, 1, 0, 1, 0, 0]
actual_jul = [0, 0, 0, 0, 0, 0, 0, 0, 0, 0]

acc_jan = accuracy_score(actual_jan, predictions)
acc_feb = accuracy_score(actual_feb, predictions)
acc_mar = accuracy_score(actual_mar, predictions)
acc_apr = accuracy_score(actual_apr, predictions)
acc_may = accuracy_score(actual_may, predictions)
acc_jun = accuracy_score(actual_jun, predictions)
acc_jul = accuracy_score(actual_jul, predictions)

errors = np.array([acc_jan, acc_feb, acc_mar, acc_apr,
acc_may, acc_jun, acc_jul])
plt.plot(errors)
plt.ylabel("Accuracy")
plt.xlabel("Index of Month")
plt.show()
```

We get a trend of accuracy dropping from the preceding code. The trend line is shown in *Figure 4.8*:

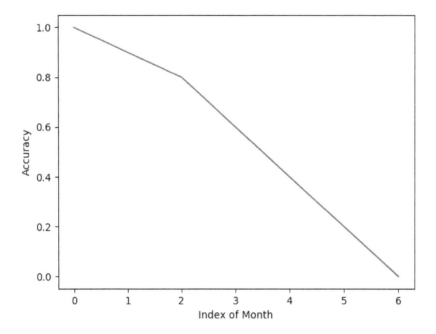

Figure 4.8 – Trend line of accuracy dropping from January 2021 to July 2022

Here, we can see that the accuracy is dropping linearly. We need to monitor the model to catch this performance drop. Otherwise, we might keep trusting the model until a big crash is exposed.

Now, let's consider a case of product recommendation. Let's assume there is a loyal customer of an e-commerce site who always trusts the product recommendations made by the site for them. Now, suddenly, the site has stopped updating the model. The loyal customer will soon find out that their favorite site is not recommending them the latest and upgraded products. Maybe the customer likes to buy the latest technical books. However, they are getting recommendations for 6-month-old books by using old technology. So, the loyal customer will quit the site and write bad reviews for it. Therefore, recommendation models need to be updated very frequently to recommend the latest products to clients based on their interests. We need to continuously monitor the model, and also the data, in this case, to upgrade the model. We need to monitor whether the model is recommending old products even though an upgraded version of the same product is already available.

In this section, we have seen some example cases to understand what can happen if a model is not evaluated continuously. We picked basic examples from different domains to demonstrate how important continuous evaluation is in those cases. In the next section, we will discuss strategies to evaluate a model continuously, along with examples.

Evaluating a model continuously

We can detect model performance in multiple ways. Some of them include:

- Comparing the performance drops using some metrics with the predictions and ground truths
- Comparing the input feature and output distributions of the training dataset are compared with the input feature and output distributions during the prediction

As an example demonstration, we will assess the model performance by comparing the predictions against the ground truths using the metrics. In this approach, to evaluate a model continuously for model performance, we have the challenge of getting the ground truth. Therefore, a major step in continuous evaluation is to collect the ground truth. So, after a model has been deployed, we need to take the following steps to continuously evaluate the model's performance:

1. Collect the ground truth.
2. Plot the metrics on a live dashboard.
3. Select the threshold for the metric.
4. If the metric value crosses the threshold limit, notify the team.

Let's look at these steps in more detail.

Collecting the ground truth

After the model has been served, we need to monitor the model performance metric that was used during training. However, to compute this metric, we need two arrays. One array will represent the predicted value and the other array will represent the corresponding actual value. However, we usually do not know the actual value of the predicted value during the time of prediction. So, what we can do is store a sample of the input and the corresponding prediction and take help from human effort to identify the correct values for those. There are multiple other ways to collect ground truths, both with human efforts and without human effort. Some of these efforts are stated here: `https://nlathia.github.io/2022/03/Labelled-data.html`. Collecting ground truth through human effort can be done in the following ways:

- *Humans can label without domain knowledge*: In these kinds of scenarios, a human can identify the actual label instantly. For example, consider a model for classifying cats and dogs. A human labeler could identify the correct label just by looking and no domain knowledge would be needed.
- *Humans with domain knowledge can label*: In this scenario, the human needs domain knowledge to identify the correct label. For example, let's consider a model that detects whether a particular disease is present or not based on symptoms. Only doctors with adequate knowledge can label this kind of data correctly. Similarly, consider identifying the variety of apples using a model. Only an agriculture expert or farmer with adequate knowledge of the varieties of apples would be able to identify the labels correctly.

- *Needing to wait to get the ground truth*: Sometimes, to know the actual value or label, we need to wait until the value or label is available. For example, consider a model predicting the price of a house. We will know the value once the house is sold. Consider another model determining whether a person can get a loan or not. We can only identify the label after the loan application has been approved or denied.

- *Needing data scraping*: You may need to scrape data from different public sources to know the actual value. For example, to know the public sentiment, know the actual price of stocks, and so on, we can take the help of web scrapers and other data-collecting tools and techniques to collect the ground truth.

- *Needing tool help*: Sometimes, the label of a feature set needs to be confirmed by a tool. For example, let's say that we have a model that recognizes the name of an acid. We need the help of tools and experiments to get the actual name of the acid.

As an example, to demonstrate saving some training examples and then collecting the ground truth later, let's say we have predictions being made against some input, as shown in the table in *Figure 4.9*, which was taken from *Figure 4.4*. This time, while making predictions, we assign the `Save?` label with `Yes`/`No` values to denote whether we should save the example, to collect the ground truth:

#Bedroom	#Bathroom	Size	Lot Size (Acre)	Prediction in Jan 2021	Save?
3	3	2000	.2	350k	Yes
4	3	2200	.3	430k	Yes
5	4	3000	.4	550k	Yes
2	2	1800	.2	300k	No

Figure 4.9 – We randomly select the input features to save to collect the ground truth

Let's assume that, for the data in *Figure 4.9*, we have decided to save 75% of the data for collecting ground truth manually. Now, we can assign a human to collect the market price of the saved houses. We will provide the features so that the housing market expert can make an educated guess or can get the price from the actual house price listed during selling. Now, the human will collect the ground truth for the saved examples, as shown in the table in *Figure 4.10*:

#Bedroom	#Bathroom	Size	Lot Size (Acre)	Price in Feb 2021
3	3	2000	.2	360k
4	3	2200	.3	445k
5	4	3000	.4	570k

Figure 4.10 – Ground truths for the saved instances are collected from the market for Feb 2021

Let's assume the ground truth collection for this model is done at the beginning of every month. So, at the beginning of March 2021, we will save another 75% of examples along with their predictions. This way, the ground truth can continue to be collected to monitor the model continuously.

Plotting metrics on a dashboard

We should have a monitoring dashboard to monitor the trend of different metrics over time. This dashboard can have multiple metrics related to monitoring model performance, along with operation metrics and business metrics. We can have a separate service just to monitor the model. Inside that monitoring service, we can fetch the recent data of predicted and actual values and plot the charts. Those charts only show recent data as these monitoring charts do not need to show a lot of history. We do not have to visualize the whole history. On the dashboard, we can create separate sections for operational metrics, business metrics, and model metrics. However, to make model retraining decisions, we usually use model metrics.

For example, let's assume that while using the instances saved for collecting ground truth, as shown in *Figure 4.10*, we get the metrics in different months, as shown in the following code snippet. Here, we are also creating a basic dashboard using the metrics:

```
import matplotlib.pyplot as plt
import numpy as np
import math

MSE = np.array([0, 4000, 7000, 11000, 15000, 23000])
MAE = np.array([0, 200, 300, 500, 700, 1000])
MAPE = np.array([0, 3, 8, 12, 20, 30])
RMSE = np.array([0, math.sqrt(4000), math.sqrt(7000),
                    math.sqrt(11000), math.sqrt(15000),
math.sqrt(23000)])

fig, ax = plt.subplots(2, 2)
fig.suptitle("Model monitoring dashboard")
ax[0, 0].plot(MSE)
ax[0, 0].set(xlabel="Index of Month", ylabel="MSE")

ax[0, 1].plot(MAE)
ax[0, 1].set(xlabel="Index of Month", ylabel="MAE")

ax[1, 0].plot(MAPE)
ax[1, 0].set(xlabel="Index of Month", ylabel="MAPE")

ax[1, 1].plot(RMSE)
ax[1, 1].set(xlabel="Index of Month", ylabel="RMSE")
```

```
plt.show()
```

The dashboard that is created from this code is shown in *Figure 4.11*:

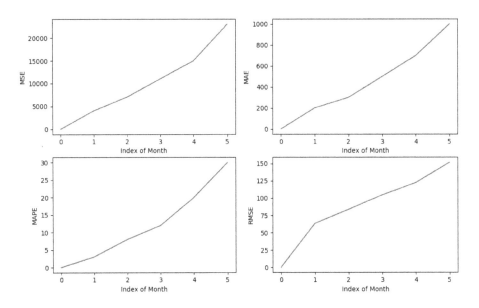

Figure 4.11 – A basic dashboard of model metrics created using matplotlib

This dashboard is very basic. We can develop more interactive dashboards using different libraries and frameworks such as `seaborn`, `plotly`, Grafana, AWS Cloudwatch, Tableau, and others.

Selecting the threshold

We can also set the threshold of the model performance metrics. The service will notify the developers when the performance starts dropping beyond the threshold. The threshold usually depends on the business goal. Some business goals will tolerate a maximum 1% drop in performance from the original performance of the model while other business goals might be okay with tolerating more of a drop. Once a threshold has been selected, then we are ready to set it in our monitoring code.

After the threshold has been selected, we can show that threshold line on the monitoring chart. For example, let's assume that we have selected an MSE threshold of 5,000 for the preceding problem. We can add the following code snippet to show a threshold line for MSE:

```
MSE_Thresholds = np.array([5000, 5000, 5000, 5000, 5000, 5000])

fig, ax = plt.subplots(2, 2)
fig.suptitle("Model monitoring dashboard")
ax[0, 0].plot(MSE)
ax[0, 0].plot(MSE_Thresholds, 'r--', label="MSE Threshold")
ax[0, 0].legend()
ax[0, 0].set(xlabel="Index of Month", ylabel="MSE")
```

Now, the dashboard will look as shown in *Figure 4.12*:

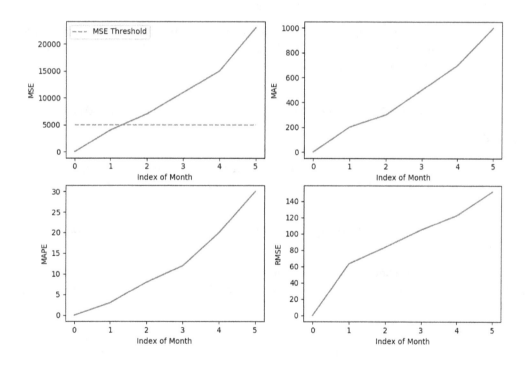

Figure 4.12 – The threshold line is shown in the chart on the dashboard for monitoring

Notice that a red dashed line is added to the MSE chart and represents the threshold for MSE. From the monitoring dashboard, we can now monitor that the MSE is approaching the threshold in February 2021. So, the engineers can plan and prepare to retrain the model.

Setting a notification for performance drops

Once the monitoring threshold has been set, we can set appropriate actions if those thresholds are crossed.

We can set an alarm and notify the developers that the model is performing below the threshold level. We can also keep track of the value of the last index in our metric array to notify customers.

For example, in our dashboard program, we can add the following code snippet to print a warning if the selected MSE threshold is crossed by the last evaluated MSE from the predictions:

```
mse_threshold = 5000
if MSE[-1] >= mse_threshold:
    print("Performance has dropped beyond the accepted
threshold")
```

We can also set other actions in this block instead of just `print`.

In this section, we discussed the approaches and strategies that need to be followed to continuously evaluate a model. In the next section, we will discuss monitoring rare classes, a special case of monitoring for continuous evaluation.

Monitoring model performance when predicting rare classes

So far, we have seen that monitoring a metric can tell us when to retrain our model. Now, let's assume that we have a model where a class is very rare and the model started failing to detect the class. As the class is rare, it might have very little impact on the metric.

For example, let's say a model is trained to classify three classes called `'Class A'`, `'Class B'`, and `'Class R'`. Here, `'Class R'` is a very rare class.

Let's assume the total predictions made by the model in January, February, and March are 1,000, 2,000, and 5,000, respectively. The instances in different classes and their correct prediction over these three months are shown in the table in *Figure 4.13*:

	January		February		March	
	Actual	**Predicted Correctly**	**Actual**	**Predicted Correctly**	**Actual**	**Predicted Correctly**
Class A	600	600	1300	1300	3000	3000
Class B	399	399	698	698	1998	1998
Class R	1	1	2	1	2	0
Accuracy	100%		99.95%		99.96%	

Figure 4.13 – Wrong prediction of rare classes might not impact the metric value significantly

From *Figure 4.13*, we can see that even though all the rare classes are being predicted wrong in March, the accuracy is shown as better than in February. So, we might think our model has very good performance, but the model has become blind to recognizing classes. How can we prevent this? Well, one simple strategy, in this case, is to assign more weight to rare classes. The value of this weight can be determined depending on how rare a class is. For our use case, we can see that the rare class appears less than once in every 1,000 examples. So, we can assume a basic metric such as the following:

```
(total_predictions/(total_classes* total_times_rare_class_
comes))
```

Therefore, the weight in January will be `(1000/(3*1))` ~333, in February will be `(2000/(3*2))` ~ 333, and in March will be `(5000/6)` ~ 833.

Now, let's define the formula to compute the weighted accuracy:

```
(total_predictions - total_wrong_predictions*weight)/total_
predictions
```

This formula is the same as the formula for computing normal accuracy with weight = 1.

Therefore, we get the modified weighted accuracy in January, February, and March as follows:

```
weighted_accuracy_Jan = (1000 - 0*333)/1000 = 100%
weighted_accuracy_Feb = (2000 - 1*333)/2000 = 83.35%
weighted_accuracy_Mar = (5000 - 2*833)/5000 = 66.68%
```

Now, we can decide that the model started performing very badly in March. Its weighted accuracy dropped to 66.68%. Therefore, we know that the model is no longer good and some actions need to be taken immediately.

In this section, we have discussed how can we use weighted metrics to monitor rare classes during continuous model evaluation. We will conclude with a summary in the next section.

Summary

In this chapter, we discussed another new pattern of model serving: the pattern of continuous model evaluation. This pattern should be followed to serve any model to understand the operational health, business impact, and performance drops of the model throughout time. A model will not perform the same as time goes on. Slowly, the performance will drop as unseen data not used to train the model will keep growing, along with a few other reasons. Therefore, it is essential to monitor the model's performance continuously and have a dashboard to enable easier monitoring of the metrics.

We have seen what the challenges are in continuous model evaluation and why continuous model evaluation is needed, along with examples. We have also looked at use cases demonstrating how model evaluation can help keep the model up to date by enabling continuous evaluation through monitoring.

Furthermore, we saw the steps that need to be followed to monitor the model and showed examples with dummy data, and then created a basic dashboard showing the metrics on the dashboard.

In the next chapter, we will talk about keyed prediction, the last model serving pattern based on serving philosophies. We will discuss what keyed prediction is and why it is important, along with techniques that you can use to perform keyed prediction.

Further reading

You can read more about metrics and continuous monitoring by looking at the following resources:

- To learn more about built-in monitoring of operation metrics supported by AWS, you can follow the resources here: `https://docs.aws.amazon.com/AmazonCloudWatch/latest/monitoring/working_with_metrics.html`

- To learn more about model performance metrics, go to `https://scikit-learn.org/stable/modules/model_evaluation.html`

5

Keyed Prediction

In this chapter, we will discuss **keyed prediction**, a pattern of model serving that involves passing a key with every training instance to facilitate an easier mapping of input to output during the collection of responses. When we need to make a large batch of predictions in the same call and multiple parallel computing servers are assigned to predict a subset of the batch, then we might lose the order of prediction. This will cause problems in mapping the prediction to the features. To solve this problem in distributed or multi-threaded serving, we pass a key along with the feature. The servers add this key along with the prediction. We can use this key to map features to predictions. In this chapter, we will discuss in detail what keyed prediction is, why it is needed, and some ways we can use keyed prediction while serving our models.

We will discuss the following topics:

- Introducing keyed prediction
- Exploring keyed prediction use cases
- Exploring techniques of keyed prediction

Technical requirements

In this chapter, we will need to have access to a Python IDE and the code from *Chapter 3* and *Chapter 4*. All this code can be found in the GitHub repository in the respective folders. The link to the *Chapter 4* codes is `https://github.com/PacktPublishing/Machine-Learning-Model-Serving-Patterns-and-Best-Practices/tree/main/Chapter%204`.

You do not have to install any additional libraries for this chapter. However, if you have difficulty accessing any library, make sure to install that library using the `pip3` command.

All the code for *Chapter 5* is present in the GitHub repository for the book under the `Chapter 5` folder. The link to the chapter's code is `https://github.com/PacktPublishing/Machine-Learning-Model-Serving-Patterns-and-Best-Practices/tree/main/Chapter%205`.

Introducing keyed prediction

When we make batch predictions of multiple instances at the same time, we split the prediction input and ask multiple servers or multiple instances of the same model to make the prediction in parallel. I hope you have already asked yourself, *if predictions are made concurrently, then how can we ensure an order*? We know that ordering is difficult in concurrent programming and the order of predictions may be violated. This creates problems on the client side. Clients may be misguided when mapping predictions to features. As multiple servers running in parallel might give the responses in different orders, it is possible that the input instances passed will be returned in a different order.

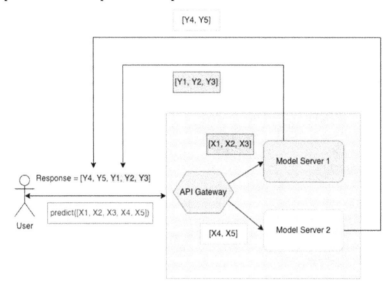

Figure 5.1 – Having parallel model servers can cause a loss of order during prediction

For example, let's consider the scenario shown in *Figure 5.1*. Here, we have two models in two servers, **Model Server 1** and **Model Server 2**, running in parallel. The API gateway splits the input data between these two models for prediction. The response returned to the client might put the response of **Model Server 2** first and the response of **Model Server 1** second because of the processing times of the models.

Although there are strategies we could use to attempt to reorder the response from the server, we still might have some failures – and if this sophisticated ordering strategy is not applied in a parallel serving infrastructure, we might get an arbitrary order in our response.

For example, in *Figure 5.1*, for the input data, [X1, X2, X3, X4, X5], we have a [Y3, Y4, Y5, Y1, Y2] response because the response from **Model Server 2**, which was assigned to predict [X4, X5], finished first and was pushed earlier to the response array. This jumbled-up order of responses can make it difficult to map a prediction to the actual input passed. To resolve this problem, we pass a key along with the prediction so that we can match the input with the prediction later on.

#Bedroom	#Bathroom	Price
5	3	500k
4	3	450k
4	2	400k
3	3	350k
3	2	300k

Figure 5.2 – Dummy house features and their prices

For example, let's consider the data for house price prediction in the table in *Figure 5.2*. During prediction, we only pass the features (#Bedroom and #Bathroom). If the predictions were made in a non-parallel environment by a single machine, then we would get the response shown in the third column, so in an ideal situation, we would get [500, 450, 400, 350, 300] as a response.

The **Mean Square Error** (MSE) in this case would be as follows:

```
MSE = [(500-500)^2 + (450-450)^2 + (400-400)^2 + (350 - 350)^2
+ (300 - 300)^2]/5 = 0
```

However, let's assume that the five features are being predicted by five different models concurrently and we get a jumbled-up response, such as [300, 350, 500, 400, 450]. As the client, we may think that we got the response in the order that we passed the features: [[5, 3], [4, 3], [4, 2], [3, 3], [3, 2]].

This time, the MSE of the prediction will not be 0 anymore. The MSE is this:

```
MSE = [(500-300)^2 + (450-350)^2 + (400-500)^2 + (350 - 400)^2
+ (300 - 450)^2]/5 = 17000
```

So, it seems that the model is performing very badly on the client side due to the misinterpretation of the order of the responses by the client.

Keyed prediction comes into the picture to solve this ordering issue and easily rearrange the mapping using keys. We pass a key along with the input feature when we make the call for the prediction. The modified table of features will look like this:

Key	#Bedroom	#Bathroom	Price
k1	5	3	500k
k2	4	3	450k
k3	4	2	400k
k4	3	3	350k
k5	3	2	300k

Figure 5.3 – Passing keys along with features during prediction

As shown in the table in *Figure 5.3*, we pass the features with a key tag. Now, the features sent to the inference server will look as follows: `[[k1, 5, 3], [k2, 4, 3], [k3, 4, 2], [k4, 3, 3], [k5, 3, 2]]`.

When we get the response, the model adds the key tag to the prediction result. This time, we will get a response such as `[(k5, 300), (k4, 350), (k1, 500), (k3, 400), (k2, 450)]`.

Even though the order is jumbled up, we now have additional information to rearrange the mapping of input to the predictions.

In this section, we have been introduced to keyed prediction. We have discussed the high-level concept of when keyed prediction is needed, along with examples. In the next section, we will discuss in detail along with code why keyed prediction is necessary. We will discuss some cases along with examples to demonstrate why the prediction will fail on the client side if we don't follow the keyed prediction pattern.

Exploring keyed prediction use cases

In this section, we will discuss the necessity of keyed prediction. First of all, we want to demonstrate some cases along with code when the input prediction mapping can be jumbled up. Then, we will discuss some applications of using keyed prediction.

There are many cases when the input/output mapping can be jumbled up. Some of these cases are as follows:

- **Multi-threaded programming**: We use multiple threads during prediction using the same machine and the threads finish in an arbitrary order

- **Multiple instances of the same model running asynchronously**: As the models run asynchronously, they may finish in any order, giving rise to the jumbling up of the prediction order

Let's discuss these two cases along with examples in the following subsections.

Multi-threaded programming

To get faster responses, we take advantage of multi-threaded programming. To make machine learning predictions, we can use multi-threaded programming to reduce the latency of the responses. However, during multi-threaded programming, the threads may finish in an arbitrary order. If we use multiple threads, then the response that comes to us may be jumbled up, and getting a one-to-one mapping from the input to the output will be difficult.

For example, let's create a dummy multi-threaded prediction scenario in the following code snippets. We have a service for predicting the data as follows. The `predict` API is exposed to the client. We also have a common utility function to combine the results, as shown in the following code block:

```
def predict(data):
    size = len(data) // 2
```

```
    thread1 = Predictor("Predictor 1", data[:size], 0)
    thread2 = Predictor("Predictor 2", data[size:], size)
    global result
    result = []
    threads = [thread1, thread2]
    thread1.start()
    thread2.start()
    for thread in threads:
        thread.join()
    return result


def combine_result(response):
    result.append(response)
```

The `predict` function is called by the client to make a prediction. The `combine_result` function is used to combine the results of parallel threads. We have used a `global result` variable to store the predictions for simplicity.

In the following part of the code, we implement a multi-threaded `Predictor` class that will be used to predict in parallel using multiple threads. The multi-threaded program is running a mock model which simply returns a pre-defined truth value. We can replace this part by adding sophisticated trained models:

```
import threading
import random
import time

threadLock = threading.Lock()

class Predictor(threading.Thread):

    def __init__(self, threadName, data, start_index):
        threading.Thread.__init__(self)
        self.threadName = threadName
        self.data = data
        self.labels = []
        self.start_index = start_index
        self.truth = [10, 11, 12, 13]
```

```
    def run(self) -> None:
        print("Predicting labels for ", self.data)
        sleepTime = random.randint(0, 10)
        print("Sleeping for ", sleepTime, " s\n")
        time.sleep(sleepTime)
        threadLock.acquire()
        self.labels = self.truth[self.start_index:self.start_
index + 2]
        combine_result(self.labels)
        threadLock.release()
        print("Predicting done:", self.data, " are predicted as
", self.labels)
```

The following code represents the client end code that is used to send requests to the server and get prediction results:

```
if __name__ == "__main__":
    data = [[1], [2], [3], [4]]
    pred = predict(data)
    print("Prediction is ", pred)
```

In the preceding code snippets, we have a `Predictor` class, which is extending the `Threading` class. That means this predictor can run in multiple threads at the same time. It has a method called `run()` that will run each time we start a thread. This method is not making predictions using any model. It is merely returning a mapping of input `[[1], [2], [3], [4]]` to `[10, 11, 12, 13]`. If we have a single thread, we will get the exact response of `[10, 11, 12, 13]`. However, in our program, we are running two threads, `thread1` and `thread2`. The two threads are assigned to provide mapping for `[[1], [2]]` and `[[3], [4]]` respectively. As no complicated logic is happening inside the threads, we have made each thread sleep for a random amount of time to demonstrate the jumbling up of the responses. Without this, we might not see the jumbling up often, as the `run()` method is not doing anything complex. In real models, a lot of complex logic will occur, often causing the threads to finish in random order.

To see how the output may be jumbled up, let's run the program twice.

Run 1

For the first run, we get the following response:

```
Predicting labels for  [[1], [2]]
```

```
Sleeping for  1  s
Predicting labels for   [[3], [4]]
Sleeping for  4  s
Predicting done: [[1], [2]]  are predicted as   [10, 11]
Predicting done: [[3], [4]]  are predicted as   [12, 13]
Prediction is  [[10, 11], [12, 13]]
```

This response looks fine to us. We would get a similar highlighted response in a single-threaded environment. Let's run the program again.

Run 2

Here's the second run:

```
Predicting labels for   [[1], [2]]
Sleeping for  3  s
Predicting labels for   [[3], [4]]
Sleeping for  1  s
Predicting done: [[3], [4]]  are predicted as   [12, 13]
Predicting done: [[1], [2]]  are predicted as   [10, 11]
Prediction is  [[12, 13], [10, 11]]
```

Now, we can see that the responses are jumbled up. This time, thread2 has finished before thread1, so the response of thread2 is appended to the output array before the response from thread1.

In this case, we have only two threads. Consider having hundreds of threads. The response will be very jumbled. For simplicity, let's consider there are a total of n threads and each of the n different threads get only one input to predict. The different number of permutations of the response is $n!$.

For example, let's suppose three threads are running, the input data is [[X1], [X2], [X3]], and the actual response from them is [Y1, Y2, Y3]. Let's also assume each of the threads is assigned to only one input instance to predict, so the number of different response orders is 3! = 3*2*1 = 6.

The responses can come in any of the following six orders:

```
[Y1, Y2, Y3]
[Y1, Y3, Y2]
[Y2, Y1, Y3]
[Y2, Y3, Y1]
[Y3, Y1, Y2]
[Y3, Y2, Y1]
```

We notice that mapping input to output is difficult and can be misinterpreted. Think about the case of continued model evaluation. For continuous model evaluation, as discussed in the last chapter, we have collected the actual value of the input, and then when we compute the metric, we get a lower value by mistake, so we will unnecessarily keep training the model, wasting our computing power, money, and time.

> **Note**
> You might have to run the preceding program a few times to get the opposite responses shown in *Run 1* and *Run 2*, as the ordering at the end of the thread is not certain. It may finish in the order of `thread1 -> thread2`, giving correct responses from time to time.

Multiple instances of the model running asynchronously

In the last section, we got predictions from the same model using multiple threads. We can also copy the same model to multiple servers and get predictions from the servers in parallel.

As the model servers run in parallel and one can finish before the other, we might get the response in a jumbled-up manner.

For example, let's have a look at the following code example. You can find the code in the GitHub repository for this chapter in the `Chapter 5` folder, which we linked in the *Technical requirements* section.

First of all, let's create a dummy model as shown in the following code block. It returns the prediction from a fixed array of truth values. It takes an index and an array and returns predictions from the `truths` array using the index:

```
truths = [10, 11, 12, 13]
import asyncio
import random


class Model:
    def __init__(self, name):
        self.name = name

    def predict(self, X, index):
        n = len(X)
        Y = truths[index: (index + n)]
        return Y
```

Then, we create an asynchronous prediction method. This asynchronous prediction method will be used by the server to make predictions from the previous model:

```
async def predict(modelName, data, model: Model, response):
    sleepTime = random.randint(0, 10)
    await asyncio.sleep(sleepTime)
    (x, i) = data
    print(f"Prediction by {modelName} for: {x}")
    y = model.predict(x, i)
    print(f"The response for {x} is {y}")
    await response.put(y)
```

The following code snippet shows the server code that calls the `async def predict` method shown previously, creates four instances of the models, and makes predictions from those four instances asynchronously. Then, we create non-blocking prediction tasks, using `asyncio.create_task`. These predictions will then run in parallel without blocking each other. For simplicity, we only predict a single example from the hardcoded data, `X`, for each of these instances.

This `server` method is called by the client for making predictions:

```
async def server():
    X = [[1], [2], [3], [4]]
    responses = asyncio.Queue()
    model1 = Model("Model 1")
    model2 = Model("Model 2")
    model3 = Model("Model 3")
    model4 = Model("Model 4")
    await asyncio.gather(
        asyncio.create_task(predict("Model 1", (X[0], 0),
model1, responses)),
        asyncio.create_task(predict("Model 2", (X[1], 1),
model2, responses)),
        asyncio.create_task(predict("Model 3", (X[2], 2),
model3, responses)),
        asyncio.create_task(predict("Model 4", (X[3], 3),
model4, responses))
    )
    print(responses)
```

The following code shows the client code that calls the server for making predictions. We are not passing any data here, as we are using some hardcoded data in the server. However, you will be passing the data when you make a call to the server during the deployment of the production model:

```
if __name__ == "__main__":
    asyncio.run(server())
```

In this example, we have a dummy prediction scenario for an input of `[[1], [2], [3], [4]]` and the truth values for this input are `[10, 11, 12, 13]`. However, this time, the predictions are being made by a clone of the same model created using the `Model` class. The models are `model1`, `model2`, `model3`, and `model4`. The `server` method here works as an API gateway in this case. It takes the client's HTTP requests and divides the prediction task between the four models.

> **API gateway**
>
> An API gateway is a service that provides API endpoints to clients and acts as a communication bridge between the backend service and the clients.

We try to simulate a non-blocking asynchronous I/O scenario to enable the servers to run in parallel in this case. To enable non-blocking prediction, we use the `Python async` keyword in the `server` function, as well as the `predict` method. Then, we create four non-blocking prediction tasks using the `asyncio.create_task` method.

> **Non-blocking operations**
>
> **Non-blocking operations** are the opposite of blocking operations. In blocking operations, the operations run sequentially, and an operation can not be started until the completion of the earlier operation. However, in non-blocking operations, the operations can run asynchronously and the execution of one operation does not impact or depend on the completion of other operations.

We store the response in a queue that the client will eventually see. Inside the `predict` method, as there is no complex logic in this dummy example, we add a random sleep to show how the response can be jumbled up.

Now, let's run the program a few times.

Run 1

Let's run it:

```
Prediction by Model 3 for: [3]
The response for [3] is [12]
Prediction by Model 4 for: [4]
The response for [4] is [13]
```

```
Prediction by Model 1 for: [1]
The response for [1] is [10]
Prediction by Model 2 for: [2]
The response for [2] is [11]
<Queue maxsize=0 _queue=[[12], [13], [10], [11]] tasks=4>
```

Here, the model servers finished in the order of model3 -> model4 -> model1 -> model2.

We see that the responses are jumbled up and do not come in the expected order of [10, 11, 12, 13].

Run 2

Let's run it a second time:

```
Prediction by Model 3 for: [3]
The response for [3] is [12]
Prediction by Model 2 for: [2]
The response for [2] is [11]
Prediction by Model 4 for: [4]
The response for [4] is [13]
Prediction by Model 1 for: [1]
The response for [1] is [10]
<Queue maxsize=0 _queue=[[12], [11], [13], [10]] tasks=4>
```

This time, the model servers have finished in the order of model3 -> model2 -> model4 -> model1. Therefore, we see that the responses can come in any order.

We have seen some examples of how the order of prediction can be lost when getting predictions from asynchronous parallel servers, or a multi-threaded computing platform. These situations clearly show that using keys would help reorder the predictions.

Why the keyed prediction model is needed

We have seen examples of jumbled-up ordering during prediction. We have an idea from these examples that keyed prediction could be used to resolve these situations.

Some of the reasons why a keyed prediction model is needed are as follows:

- To rearrange the order for the client
- To compute correct metrics during continued model evaluation
- To store the predictions separately from the features

Rearranging the order of prediction

The keyed prediction model is needed to reorder the predictions in situations when the same ordering of input and output is not guaranteed. For example, let's consider the earlier cases, where we used the example input of [[1], [2], [3], [4]] and the corresponding expected response is [10, 11, 12, 13]. We have shown that the response, [10, 11, 12, 13], may be jumbled up before it is returned to the client. Now, if we pass keys along with the features in this way, [[k1, 1], [k2, 2], [k3, 3], [k4, 4]], then we can return these keys along with the predictions during prediction. For example, the prediction for [k1, 1] will now be returned as [k1, 10] instead of just 10, so from a response of [k1, 10], we can now tell that this response corresponds to the [k1, 1] input. This way, we can identify the exact mapping between the input and the prediction to reorder the responses.

For example, let's have a look at the following code snippet:

```
import random
truths = [10, 11, 12, 13]

def jumbled_prediction_without_keys(X):
    response = []
    for i in range(len(X)):
        response.append(truths[i])
    random.shuffle(response)
    return response

def jumbled_prediction_with_keys(X):
    response = []
    for i in range(len(X)):
        (key, x)   = X[i]
        response.append((key, truths[i]))
    random.shuffle(response)
    return response


if __name__ == "__main__":
```

```
X1 = [[1], [2], [3], [4]]
Y1 = jumbled_prediction_without_keys(X1)
print("Predictions without keys ", Y1)
X2 = [(0, 1), (1, 2), (2, 3), (3, 4)]
Y2 = jumbled_prediction_with_keys(X2)
print("Predictions with keys ", Y2)
Y2.sort(key=lambda pred: pred[0])
print("Ordering restored using keys ", Y2)
```

From this code snippet, we get the following output:

```
Predictions without keys  [12, 11, 10, 13]
Predictions with keys  [(3, 13), (1, 11), (0, 10), (2, 12)]
Ordering restored using keys  [(0, 10), (1, 11), (2, 12), (3,
13)]
```

We have two functions for making some dummy predictions. In one function, `jumbled_prediction_without_keys`, we do not pass the keys and the response is shuffled in an arbitrary order, so we get a response that cannot be restored in order.

From the output, we can see that the predictions are `[12, 11, 10, 13]`. We do not have additional information to map this response to the input.

In the other function, `jumbled_prediction_with_keys`, we pass keys along with the input and also tag the predictions with those keys. In the example, we passed the input as `[(0, 1), (1, 2), (2, 3), (3, 4)]`. Here, 0, 1, 2, and 3 are the keys. From the function, we got a response of `[(3, 13), (1, 11), (0, 10), (2, 12)]`. Although the responses are jumbled up, we can now map the input to the predictions easily. For example, in the code, we have rearranged the response using keys in `Y2.sort(key=lambda pred: pred[0])`.

Computing the correct metrics during continued evaluation

Let's recall the process of continuous model evaluation from the last chapter. We save a sample of predictions, then we collect the ground truth for the predictions, and after that, we compute the value of the evaluation metric using the truth and the predicted values. We will keep collecting the ground truth for the features in the order that the features are saved. However, if the predictions are jumbled up, we will compute the metric incorrectly.

For example, let's say that we have the following feature table that is passed as input to the model for prediction:

F1	F2	Prediction
x1	y1	100
x2	y2	500
x3	y3	1000
x4	y4	1500
x5	y5	2000

Figure 5.4 – A feature table with actual prediction values

Let's assume we save the $[(x2, y2), (x4, y4)]$ features at indices 1 and 3 for continued model evaluation. The model has returned predictions in an order of $[500, 2000, 1000, 100, 1500]$. Therefore, for continuous evaluation, we assume that the predicted values for $[(x2, y2), (x4, y4)]$ are $[2000, 100]$. Now, let's say the ground truth values we collect from the recent sales data are $[501, 1501]$ for $[(x2, y2), (x4, y4)]$, which are very close to the predictions made by the model. However, because of the jumbled-up prediction order, the continuous model evaluator will think that the model has predicted 2000 instead of 501, and 100 instead of 1501. The continuous model evaluator will decide that the model is performing much below the accepted threshold and will provide false alarms to the developer team. To refresh your memory on continuous model evaluation techniques, refer to *Chapter 4*.

Storing features and predictions separately

When we have large-scale data coming in periodically, we might prefer to store the prediction results in a separate table. Now, the question is how we can separate these tables and, later on, match the two tables. If you have a background in working with SQL databases, you will know that we can use a key that can help us to join the records later on.

For example, let's use MySQL to create two tables, `Features` and `Prediction`. We can create dummy tables and insert some dummy data using the following query:

```
CREATE TABLE IF NOT EXISTS `Features` (
    `key` varchar(50) NOT NULL,
    `F1` int(6) NOT NULL,
    `F2` int(6) NOT NULL,
    PRIMARY KEY (`key`)
) DEFAULT CHARSET=utf8;
```

```
INSERT INTO `Features` (`key`, `F1`, `F2`) VALUES
  ('k1', '3', '3'),
  ('k2', '5', '4'),
  ('k3', '4', '3'),
  ('k4', '4', '2');


CREATE TABLE IF NOT EXISTS `Predictions` (
  `key` varchar(50) NOT NULL,
  `Price` int(8) NOT NULL
) DEFAULT CHARSET=utf8;
INSERT INTO `Predictions` (`key`, `price`) VALUES
  ('k2', '500'),
  ('k4', '350'),
  ('k3', '400'),
  ('k1', '300');
```

We can test the queries online at https://www.db-fiddle.com. I have already created DB fiddle for you at https://www.db-fiddle.com/f/4pyrD3WoUqg8MWoCEkMnXB/0. The advantage of using DB Fiddle is that you do not need to create tables on your local computer and you also do not need to install a database query engine.

In this query, we are creating two tables, Features and Predictions. The Features table has three columns, key, F1, and F2. Here, key is the column that contains the keys we create for keyed prediction. F1 and F2 are the columns indicating two features. The Predictions table contains two columns, key and price. Here, key is the column that corresponds to the keys passed by the clients with the features, and price is the column for the predicted values.

After creating the tables, we also insert some dummy data into the tables. In the Predictions table, we insert the predictions in a different order than the features. The actual order of assumed predicted prices should be [300, 500, 400, 350]. However, in the table, we inserted them in an order of [500, 350, 400, 300].

Now, let's test the members in the table to ensure that we have the correct data. We can run the SELECT * FROM Features; query in DB Fiddle. We will see the response in *Figure 5.5*:

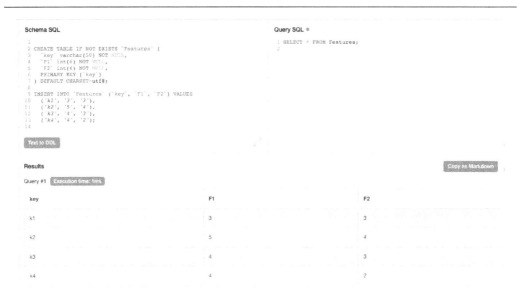

Figure 5.5 – The Features table viewed from www.db-fiddle.com

We can see that the table has been created with the correct data. The keys and **F1** and **F2** features are correctly populated.

Now, in the same way, let's run the `SELECT * FROM Predictions;` query to see the `Predictions` table. The table is shown in *Figure 5.6*:

Figure 5.6 – The Predictions table viewed from www.db-fiddle.com

We have intentionally inserted the keys, so we can see that the predicted prices are in a different order than that of the features to show that the predictions may be written in an arbitrary order.

Now, we can use the keys to easily match the predictions with the features. For example, let's match the predictions to the features in the two preceding tables.

We can write the following query to do the match:

```
SELECT f.key, f.F1, f.F2, p.price
        FROM Features f
            JOIN Predictions p
                on f.key=p.key;
```

If we run this query in DB Fiddle, we will see the Features table and the Predictions table are joined in the correct mapping of input to predictions, as shown in *Figure 5.7*:

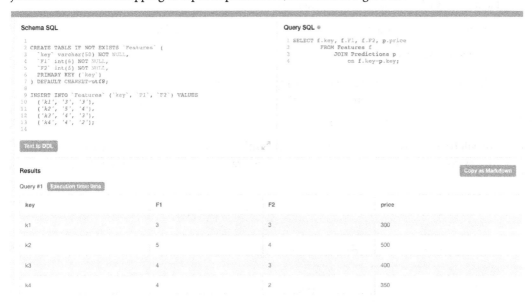

Figure 5.7 – The mapping of features and predictions using the keys

We see that by using keyed prediction pattern, we can easily maintain two separate tables for features and predictions. We can join the two tables using the keys to exactly map a feature to the correct prediction.

If we do not use keyed prediction, then we might have two options:

- Store the features along with the predictions only after the prediction is made, but now we are at the risk of losing data if some predictions fail.

- Keep a separate table where full features are stored with the predictions. This unnecessarily causes complexity in the table and unnecessarily duplicates data.

Therefore, the keyed prediction pattern gives us a smart solution for storing predictions and features separately and maintaining them as separate entities.

In this section, we have discussed the necessity of keyed prediction and demonstrated some cases where keyed prediction might be needed. In the next section, we will talk about the techniques needed for keyed prediction.

Exploring techniques for keyed prediction

In this section, we will discuss the ideas and techniques needed for keyed prediction. Keyed prediction has three main tasks that need to be performed at a high level:

- Passing keys with features from the client side
- Removing keys from the features before being passed to the model for prediction
- Tagging predictions with keys from the server side

We will discuss these three tasks in the following subsections. To describe these three steps, let's first create a basic keyed prediction scenario using `pandas` and `scikit-learn` in the following code snippet:

1. We will train a basic linear regression model with the data passed from the client. We avoid the key column during training by specifying the trainable columns in the `Features` array:

```
import pandas as pd
from sklearn.linear_model import LinearRegression


Features = ["#Bedroom", "#Bathroom"]


def train(X: pd.DataFrame, y: pd.Series):
    X = X.loc[:, Features]
    model = LinearRegression()
    model.fit(X, y)
    return model
```

2. We create functions for prediction in the following code snippet. During prediction, we append the keys along with the corresponding predicted values. We have a function to make a single prediction at a time and we pass the single input along with the key to this function. While returning the response from this function, we append the key with the response.

```python
def predict_single(model: LinearRegression, x, key):
    y = model.predict(x)
    pred = pd.DataFrame({"y": [y], "key": [key]})
    return pred


def predict_batch(model: LinearRegression, X:
pd.DataFrame):
    X = X.sample(frac=1)
    keys = X["key"]
    response = pd.DataFrame()
    for index, row in X.iterrows():
        pred = predict_single(model, [[row[Features[0]],
row[Features[1]]]], keys[index])
        response = response.append(pred, ignore_
index=True)
    return response
```

3. In the following code snippet, we have the program used by the client to call the prediction API. We see that from the client side, during prediction, we pass keys along with the input data:

```python
if __name__ == "__main__":
    X = pd.DataFrame({"#Bedroom": [5, 5, 4, 4, 3],
"#Bathroom": [4, 3, 3, 2, 2]})
    Y = pd.Series([500, 450, 400, 350, 300])
    model = train(X, Y)
    X['key'] = pd.Series(["1", "2", "3", "4", "5"])
    response = predict_batch(model, X)
    print("Final response: ")
    print(response)
```

In this code snippet, we have client code within if __name__ == "__main__". We have a hypothetical server exposing one API, predict_batch. To mimic the asynchronous parallel execution, we have shuffled the input at the beginning of the predict_batch function. The predict_single function represents each model server running in parallel. There is a function called train to train the model with some basic dummy data. While serving the model, we usually do the training before sending the model for serving. We make predictions with the same data used for prediction training to show you the relation between predictions and keys clearly. In an ideal situation, we would split the dataset before training and use different sets of data for training and testing.

From the code snippet, we see the following output:

```
                        y key
0                   [350.0]   4
1   [499.99999999999994]   1
2   [449.99999999999994]   2
3   [300.00000000000006]   5
4                   [400.0]   3
```

We have the keys column along with the predictions. The order of prediction is jumbled up, but we can now easily map the input to the output using the keys.

Passing keys with features from the clients

For keyed prediction, the point from which keys propagate is the client side. When a client passes multiple inputs as a batch for prediction, the client needs to assign a unique key to each input and pass the keys as a separate column:

1. In the pandas library, we can easily append a new column using df['key'] = [k1, k2, k3, ...].

    For example, in the preceding code snippet, the client has the following training data:

    ```
    X = pd.DataFrame({"#Bedroom": [5, 5, 4, 4, 3],
    "#Bathroom": [4, 3, 3, 2, 2]})
    Y = pd.Series([500, 450, 400, 350, 300])
    ```

2. Now, after the model is trained and served, we make predictions with the same value of X. The client needs to pass the keys before making predictions for the following keyed prediction pattern, so we use the following code snippet to assign unique keys to each input instance:

    ```
    X['key'] = pd.Series(["1", "2", "3", "4", "5"])
    ```

3. If we check the data using print(X.head()), we will see the following output:

    ```
        #Bedroom  #Bathroom key
    0          5          4   1
    1          5          3   2
    2          4          3   3
    ```

| 3 | 4 | 2 | 4 |
| 4 | 3 | 2 | 5 |

The column key is now appended with our data being passed for prediction.

This is the additional responsibility assigned to the client program during prediction. However, this burden on the client can also be removed.

4. The model deployment engineers can take responsibility for creating keys and assigning a key to each input instance. Then, after making predictions, they can restore the prediction order using the keys. In that case, in our example, the only change that is needed is to move X['key'] = pd.Series(["1", "2", "3", "4", "5"]) from within the client block to the beginning of the predict_batch function.

Everything else will remain the same. However, it has a drawback in that we will be adding additional computation burden on the server end, worsening the latency of the API.

Removing keys before the prediction

The key column is not a part of the feature in the model, so if we pass the key column along with the features, we will get an error during prediction. For example, in the preceding code snippet, let's modify the predict_batch function to pass the key column to the model along with other features. We modify the call of predict_single to pred = predict_single(model, [[row['key'], row[Features[0]], row[Features[1]]]], keys[index]):

```
def predict_batch(model: LinearRegression, X: pd.DataFrame):
    X = X.sample(frac=1)
    keys = X["key"]
    response = pd.DataFrame()
    for index, row in X.iterrows():
        pred = predict_single(model, [[row['key'],
row[Features[0]], row[Features[1]]]], keys[index])
        response = response.append(pred, ignore_index=True)
    return response
```

Now, if we run the program, we will get the following error:

```
ValueError: matmul: Input operand 1 has a mismatch in its core
dimension 0, with gufunc signature (n?,k),(k,m?)->(n?,m?) (size
2 is different from 3)
```

We're getting this error because we have trained the model with only two features and now, during prediction, we are passing three features. That's why we see an error message of (size 2 is different from 3).

Therefore, we remove the keys before sending them for prediction and we also need to send the corresponding key to the model server. We have to go through the following steps for key removal:

1. During the key removal process, we select the expected columns using the column names. We need to have a separate array indicating the feature names. In our example, we have used the Features array to store the feature names used during training.

2. Only use the features that were used during training during the time of prediction. For example, in the preceding code snippet, we select the expected two feature columns in the [[row[Features[0]], row[Features[1]]]] section of the code. We select, the Features[0] column, which means #Bedroom, and the Features[1] column, which means #Bathroom. These are the features that were used to train the model.

Tagging predictions with keys

When we make predictions for an input instance, we assign the provided key for that input as the key for the prediction as well. In the preceding example, this happens in the following code snippet:

```
def predict_single(model: LinearRegression, x, key):
    y = model.predict(x)
    pred = pd.DataFrame({"y": [y], "key": [key]})
    return pred
```

To tag the key, we create a DataFrame for the response and add the prediction and the key columns for predictions made by a single thread or single instance of the model at a time.

This helps identify the predictions and map the predictions to the exact inputs from the clients. Although we have jumbled up the DataFrame before prediction, we can now easily map the input to the predictions using the keys. For example, in the response to the preceding code, the first row is [350] 4. We can easily say that although this response appears at the top, it is not the prediction for the [5, 4] features – it is the prediction for the fourth row in the input with the [4 2] features.

So far, we have talked about the steps that should be followed during keyed prediction. In the next subsection, we will talk about different key creation ideas.

Creating keys

An important step of keyed prediction is the creation of the keys that need to be passed along with the training data because if the keys are not unique, then the keyed prediction will be ambiguous and unsuccessful.

There are multiple ways to create keys. We will discuss the following techniques:

- Using the index as the key

- Using a hash value as the key

Using the index as the key

In this approach, we use the index value of input as the key.

For example, let's consider the feature table in *Figure 5.8*. In this table, we use the index of the input as the key. The index of the first input, [300, True], is 0 and hence the key for this row is also 0.

F1	F2	key
300	True	0
500	False	1
600	False	2

Figure 5.8 – We can use the index of a feature set as the key for that feature set

This is a very simple way to create keys and would work for many use cases. However, there is a drawback to this approach. The same index will be present in all the requests. For example, suppose a user sends two prediction requests, *R1* and *R2*. In both *R1* and *R2*, there will be inputs with an index of 0, which means they will have the same key, 0. Now, think for a moment about what the issue is. If the client is collecting the responses from multiple prediction requests together, then the keyed prediction will fail because the keys will now be duplicated, creating confusion.

For example, let's say a client has sent three requests, *R1*, *R2*, and *R3*:

- The inputs for *R1* are [[m1, n1, 0], [m2, n2, 1]]

- The inputs for *R2* are [[m3, n3, 0], [m4, n4, 1]

- The inputs for *R3* are [[m5, n5, 0], [m6, n6, 1]]

All three requests have the same keys, 0 and 1. Now, the user has collected the responses from these three requests together, as shown in *Figure 5.9*:

key	prediction
0	p1
1	p2
0	p3
1	p4
0	p5
1	p6

Figure 5.9 – Using the indexes as keys can give rise to duplicates

We notice that for the same key, 0, we have three different records of predictions, so we are again failing to do one-to-one mapping of input to output.

The situation can be worse if we save predictions from a large amount of data in a prediction table daily. Every day or every hour, we make predictions and append the results to a prediction table. If we use the index as the key, the table will have many duplicate keys appearing regularly, so the prediction table will not be useful at all.

To demonstrate how problematic mapping can be if we allow duplicate keys using the indexes, let's use the following queries:

1. Create a table named Features with key, F1, and F2 columns using the following query:

    ```
    CREATE TABLE IF NOT EXISTS `Features` (
        `key` int(6) NOT NULL,
        `F1` int(6) NOT NULL,
        `F2` int(6) NOT NULL
    ) DEFAULT CHARSET=utf8;
    ```

2. Insert some dummy data into the table created in *step 1* using the following query. Two separate insert statements indicate the data inserted for two different requests. As indexes are used as keys, the keys start from 0 in each of the requests:

    ```
    INSERT INTO `Features` (`key`, `F1`, `F2`) VALUES
        (0, '3', '2'),
        (1, '5', '4');

    INSERT INTO `Features` (`key`, `F1`, `F2`) VALUES
        (0, '3', '3'),
        (1, '5', '5');
    ```

3. Create a Predictions table with two columns, key and Price, using the following query:

    ```
    CREATE TABLE IF NOT EXISTS `Predictions` (
        `key` varchar(50) NOT NULL,
        `Price` int(8) NOT NULL
    ) DEFAULT CHARSET=utf8;
    ```

4. Insert predictions data into the `Predictions` table using the following query. The two separate insertions are for two different prediction requests:

```
INSERT INTO `Predictions` (`key`, `price`) VALUES
(1, '500'),
(0, '300');

INSERT INTO `Predictions` (`key`, `price`) VALUES
(0, '320'),
(1, '550');
```

5. Join the prediction from the `Predictions` table and the input from the `Features` table to map them:

```
SELECT f.key, f.F1, f.F2, p.price
FROM Features f
     JOIN Predictions p
          on f.key=p.key
                order by f.key;
```

We enter the features and the predictions for *two* different requests in the table. Both requests have only two inputs and therefore get two responses. We can run these queries in DB Fiddle. I have created a link, `https://www.db-fiddle.com/f/9bvdfjwyoNmuuewCDgXFY5/0`, with the queries there.

6. Now, if we try to map the prediction table with the features table using the query shown in *step 5*, we get the response shown in *Figure 5.10*:

Key	F1	F2	Price
0	3	2	300
0	3	3	300
0	3	2	320
0	3	3	320
1	5	4	550
1	5	5	550
1	5	4	500
1	5	5	500

Figure 5.10 – With duplicate keys, mapping the features to the predictions becomes a challenge

For each key, there are four matches, and half of them are false. Among the matches, [0, 3, 2] -> 300 is correct and [0, 3, 3] -> 320 is correct for the 0 key. However, the other two mappings [0, 3, 2] -> 320 and [0, 3, 3] -> 300, are incorrect. As we are using the same key in the Features table and the Predictions table, these are also returned as valid matches during the creation of mapping from input to prediction using the query shown in *step 5*. [0, 3, 2] -> 320 is returned because the key for the [3, 2] feature is 0 in the Features table and the key for the 320 prediction is 0 in the Predictions table. Therefore, there is no reason for [0, 3, 2] -> 320 not to be considered a correct mapping, technically. Similarly, [0, 3, 3] -> 300 is also considered correct. The same applies to the 1 key.

Therefore, in cases when we need to append the predictions from multiple calls in a central persistence store, we will get erroneous results if we use the index as the key.

Using a hash value as the key

Another way to create keys is to use a hash value as the key. In this case, we hash the features in the input and use the hash value as the key. In an ideal situation, the hash values can not be duplicated, so we are confident that we do not have duplicate values. For example, let's consider the following code snippet:

```
import pandas as pd


df = pd.DataFrame({"F1": [3, 4, 5, 6], "F2": [100, 200, 300,
400]})
print(df.head())
df['key'] = df.apply(lambda x: hash(tuple(x)), axis=1)
print(df.head())
```

The output for this code is the following:

```
    F1    F2
0    3   100
1    4   200
2    5   300
3    6   400

    F1    F2                    key
0    3   100  -5165029896791181875
1    4   200  -2728380094227174081
2    5   300    197667696074879495
3    6   400   8793855073643805160
```

The hash values are added as keys and are visible in the response thanks to the second `print` statement, `print(df.head())`.

This approach assigns the same key to the same features. We can also avoid storing the predictions for duplicate keys if the prediction values are the same.

Using hashing techniques to generate keys in keyed prediction gives us the following advantages over using indexes as keys:

- Keys generated as hash values are unique, so we get rid of the problems discussed in this chapter caused by the presence of duplicate keys if input indexes are used as keys.

- Keys generated using hash values can help maintain big prediction pipelines, keeping prediction and feature tables separate. As the keys are unique, we can use the keys to join the tables and map inputs to predictions without any ambiguity.

Using hashing techniques to generate keys also has some disadvantages, such as the following:

- Hashing can create larger keys, requiring additional memory to store these keys. This can be a barrier for edge devices and embedded systems where memory is a big concern.

- Hashing can't provide information about the time when the key was generated, so in systems where we want to group the inferences based on a certain time window, we can also use a timestamp for generating keys. You should also consider adding a timestamp for this reason so that predictions made at certain times can be identified quickly.

In this section, we discussed different techniques in keyed prediction and we concluded the section by discussing how can we create keys. We discussed how using indexes as keys can fail if we have large-scale data, so using indexes as keys can only be applied in limited cases where we need predictions on a small scale. However, in large-scale predictions, using unique hash values as keys can be a preferable approach due to its mechanism to handle duplicates. In the dataset, if there is already a unique key being used, we also use that as a key in this pattern. For example, if we have unique IDs for items in a store that needs demand prediction, then we can use those unique IDs as keys instead of regenerating the keys. In the next section, we will conclude our chapter by drawing a summary.

Summary

In this chapter, we discussed using the keyed prediction pattern when serving models. This is an important pattern when our serving environment is making predictions on a large scale asynchronously. In this case, we risk losing the order of the predictions and we will not be able to map the inputs to the corresponding predictions. To solve this problem, we should tag the training data with keys and get predictions with the same keys tagged. In this way, we can reproduce the exact mapping of input to output.

We also discussed situations in which keyed prediction is needed and looked at some examples. Then, we discussed the techniques of keyed prediction. We concluded with some ideas for the creation of keys during keyed prediction.

In the next chapter, we will start talking about a different group of patterns for serving models. These serving patterns fall into the second category of patterns as discussed in *Chapter 2*. These patterns deal with the exact serving mechanism and the best practices to follow. We will discuss the batch model serving pattern in the next chapter.

Further reading

Consult the following resources to enhance your reading:

- Find out more about hashing at `https://www.okta.com/identity-101/hashing-algorithms/`

- Find out more about SQL joins at `https://www.okta.com/identity-101/hashing-algorithms/`

Batch Model Serving

The batch model serving pattern is more common than real-time serving in today's world of large-scale data. With batch model serving, newly arrived data will not be immediately available in the model. Depending on the batching criteria, the impact of new data may become apparent after a certain period of time, ranging from an hour to a year. Batch model serving uses a large amount of data to build a model. This gives us a more robust and accurate model. On the other hand, as the batch model does not use live data, it cannot provide up-to-date information during prediction. For example, let's consider a product recommendation model that recommends shirts, and the model is retrained at the end of every month. You will get a very accurate shirt recommendation based on the available shirts up to the last month, but if you are interested in shirts that have recently appeared on the market, this model will frustrate you, as it doesn't have information about the most recent products. Therefore, we see that during batch serving, selecting the ideal batching criteria is a research topic that can be adjusted based on customer feedback.

In this chapter, we will discuss batch model serving. After completing this chapter, you will know what batch model serving is, and you will have seen some example scenarios of batch serving. The topics that will be covered in this chapter are listed here:

- Introducing batch model serving
- Different types of batch model serving
- Example scenarios of batch model serving
- Techniques in batch model serving

Technical requirements

In this chapter, we will discuss batch model serving, along with some example code. To run the code, please go to the Chapter 6 folder in the GitHub repository for the book and try to follow along: https://github.com/PacktPublishing/Machine-Learning-Model-Serving-Patterns-and-Best-Practices/tree/main/Chapter%206.

We shall be using **cron** expressions to schedule batch jobs in this chapter. The `cron` commands that will be run in this chapter need a Unix or Mac **operating system** (**OS**). If you are using Windows, you can use a virtual machine to run a Unix OS inside the virtual machine. Use this link to learn about some virtual machines for Windows: `https://www.altaro.com/hyper-v/best-vm-windows-10/`.

Introducing batch model serving

In this section, we will introduce the batch model serving pattern and give you a high-level overview of what batch serving is and why it is beneficial. We will also discuss some example cases that illustrate when batch serving is needed.

What is batch model serving?

Batch model serving is the mechanism of serving a machine learning model in which the model is retrained periodically using the saved data from the last period, and inferences are made offline and saved for quick access later on.

We do not retrain the model immediately when the data changes or new data arrives. This does not follow the CI/CD trend in web application serving. In web application serving, every change in the code or feature triggers a new deployment through the CI/CD pipeline. This kind of continuous deployment is not possible in batch model serving. Rather, the incoming data is batched and stored in persistent storage. After a certain amount of time, we add the newly collected data to the training data for the next update of the model.

This is also known as **offline model serving**, as the model does not represent recent data and the training and inference are usually done offline. By offline, here, we mean that the training and inference are done before the predictions are made available to the client. This time gap between training and using the inference is higher than the acceptable latency requirements for online serving, as clients can't wait this long for a response after the request is made.

To understand batch model serving, first, let's discuss two metrics that are crucial to serving and also to the monitoring of the health of serving. These metrics are as follows:

- Latency
- Throughput

Latency

Latency is the difference between the time a user makes a request and the time the user gets the response. The lower the latency, the better the client satisfaction.

For example, let's say a user makes an API call at time T_1 and gets the response at time T_2.

Then, the latency is $T_2 - T_1$.

If the latency is high, then the application cannot be served online because there is a maximum timeout limit for an HTTP connection. By default, the value is set to 30 seconds and the maximum value is 5 minutes. However, in an ideal situation, we expect the response to be within milliseconds. AWS API Gateway sets a hard limit of 29 seconds before you encounter an error:

```
{
Message: "Endpoint request timed out"
}
```

Therefore, high latency discourages serving applications to the users.

Because of this latency constraint, we can understand why batch inference can't serve the most recent large volumes of data. The modeling (training and evaluation) and inference will take a lot of time, and completing everything before this HTTP timeout period expires is not possible if the data volume is high and the inference needs heavy computation. That's why in batch serving, we need to sacrifice latency and get our inference after a certain period.

Throughput

Throughput is the number of requests served per unit of time. For example, let's say 10 requests are successfully served in 2 seconds by a server. The throughput in this case is *10/2 = 5* requests per second. Therefore, we want high throughput from the served application. To achieve this, we can scale horizontally by adding more servers and using the high computation power in the servers.

From our discussion of latency and throughput, you may have noticed that during serving, the goal should be to minimize latency and maximize throughput. By minimizing latency, we ensure customers get the response quickly. If the response does not come quickly, the users will lose interest in using the application. By maximizing throughput, we ensure that more customers can be served at the same time. If throughput is low, customers will keep getting a `"Too many requests"` exception.

In batch model serving, we want to maximize the throughput so that large numbers of clients can get predictions. For example, let's say one million customers are using an e-commerce site. All of them need to get their product recommendations at the same time.

During batch serving, we give priority to throughput. We want all the clients to get their predictions at the same time to ensure the fairness of service. However, we get the inference result a certain amount of time after the inference request is sent.

Main components of batch model serving

Batch model serving has three main components:

- **Storing batch data in a persistent store**: During batch model serving, we cannot use live data to update the model. We have to wait for a certain amount of time and retrain the model using the data. For that reason, we need to store the data in a persistent store and use the data from the persistent store to train the model. Sometimes, we might want to use the data only from a certain recent period of time. The database we choose depends on the amount of data we're going to work with. It may be the case that it is sufficient to store the data in CSV files; it may also be necessary to move it to specialized databases for big data, such as Hadoop or Spark. We can organize the files or tables and separate them. For example, we can keep a separate table for each month, or we can keep a file for each genre in a movie or book recommendation model data.

- **Trigger for retraining the model**: We need to decide when the model will be retrained. This can be done manually through manual observations, it can be done automatically with scheduling, or the model can be retrained using the evaluation metric by using continuous evaluation, as discussed in *Chapter 4, Continuous Model Evaluation*.

- **Releasing the latest inference to the server**: After the model is trained and the inference is completed, we can then write the inference result to the database. However, we need to let the servers know that the update to the model is complete. Usually, we can follow a notification strategy to publish a notification with the information that the update is completed.

In this section, we have introduced batch serving to you. In the next section, we will see when to use different types of batch model serving based on different criteria. We will also introduce `cron` expressions for scheduling periodic batch jobs during batch model serving.

Different types of batch model serving

Batch model serving can be divided into a few categories with different parameters and settings. The characteristics that can create differences in batch model serving are as follows:

- Trigger mechanisms
- Batch data usage strategy
- Based on the inference strategy

We can classify batch model serving into three categories based on the trigger mechanism:

- Manual triggers
- Automatic periodic triggers
- Triggers based on continuous evaluation

Let's look at each of them.

Manual triggers

With a **manual trigger**, the developer manually starts the retraining of the model and only then does the model start retraining using the most recent saved data. The developer team can keep monitoring the performance of the model to find out when the model needs retraining. The retraining decision can also come from the business users of the model. They can request the developer team for retraining and then the developer can retrain the model.

The manual trigger has the following advantages:

- Unnecessary computation can be avoided. As computation is not free, we can save a lot of money by avoiding periodic retraining unless the clients need the latest update.

- By deploying manually, the developers can observe and fix any problems arising during the training.

However, this approach has the following disadvantages:

- There is an added burden on the developer team to monitor the model's performance and decide when retraining is needed

- Manual observation may be erroneous or biased

Let's see how to monitor this in practice.

Monitoring for manual triggers

For manual observation, the developer team might need to conduct surveys to understand whether the clients are satisfied and are getting the desired predictions. The survey questions can be designed based on the model's business goals. For example, for a product recommendation site, we can set up a survey where users rate how happy they are with their recommendations. We can then check the average rating and see that it is dropping every time. If the average rating approaches a certain threshold, the developers can retrain the model.

To see a code example of this, let's say the developer sends surveys to five users every week and the survey responses are as follows:

User	Week 1	Week 2	Week 3	Week 4	Week 5
A	5	5	4	3	3
B	5	4	4	3	3
C	5	4	4	4	3
C	5	5	4	4	3
E	5	5	4	4	3

Figure 6.1 – Rating submitted by five users over different weeks

In the table in *Figure 6.1*, the users' ratings are collected over 5 different weeks. The users submit a rating out of five based on how satisfied the users are with their product recommendations. The developers can use this result to decide when they should start retraining the model. The deciding criteria can be whatever you want. For example, the following criteria may be used:

- When the average rating is $<= 4.0$, start retraining. In this case, the developers will start retraining in week 3.

- When the first rating of $<= 3$ comes in, start retraining. In that case, the developers will retrain in week 4.

- When the median rating is $<= 3$, then do the retraining. In that case, the developers will start retraining in week 5.

These criteria are dependent on the team. However, we can see from the criteria mentioned here that the training can happen in **Week 3**, **Week 4**, or **Week 5**.

We can use these conditions listed previously as flags that can indicate whether retraining should be done or not. We can compute the previously listed mentioned triggers in the following way:

```python
import numpy as np
users = ["A", "B", "C", "D", "E"]
w1 = np.array([5, 5, 5, 5, 5])
w2 = np.array([5, 4, 4, 5, 5])
w3 = np.array([4, 4, 4, 4, 4])
w4 = np.array([3, 3, 4, 4, 4])
w5 = np.array([3, 3, 3, 3, 3])


def check_any_less_than_3(rating):
    mappedArray = rating <= 3
    return mappedArray.any()


print("Checking when first rating <= 3 appears")
print(check_any_less_than_3(w1))
print(check_any_less_than_3(w2))
print(check_any_less_than_3(w3))
print(check_any_less_than_3(w4))
print(check_any_less_than_3(w5))


def check_average(rating):
    mean = np.mean(rating)
```

```
    return mean <= 4.0

print("Checking when average rating becomes <= 4.0")
print(check_average(w1))
print(check_average(w2))
print(check_average(w3))
print(check_average(w4))
print(check_average(w5))

def check_median(rating):
    median = np.median(rating)
    return median <= 3.0

print("Checking when the median becomes <= 3.0")
print(check_median(w1))
print(check_median(w2))
print(check_median(w3))
print(check_median(w4))
print(check_median(w5))
```

From this code snippet, we get the following response:

```
Checking when first rating <= 3 appears
False
False
False
True
True
Checking when average rating becomes <= 4.0
False
False
True
True
True
Checking when the median becomes <= 3.0
False
False
```

```
False
False
True
```

In the response, we have highlighted when the trigger responds with `True`. We can see that `True` is returned in different weeks for different criteria, so if the business case is sensitive, the developers can choose the criterion that allows them to train the model as early as possible to ensure greater customer satisfaction. In that case, the developers can monitor both criteria and work on whichever comes first. If the business case does not require such tight accuracy, then the developers can use the criterion that gives them less pressure in terms of redeployment and reduce their computation costs.

Another approach to monitoring may be to keep a light model in parallel with the big model in production. The light model will be trained regularly for a limited number of randomly selected users. Then, we can monitor the inference of the model for those users in the production server and compare it with the inference made by the light model in the development server. For example, let's consider the product recommendations for a user from a batch model and from a light model, which are shown in the table in *Figure 6.2*.

Recommendation from batch model	Recommendation from light, continuously updating model	Rank
Product 3	Product 10	1
Product 4	Product 9	2
Product 1	Product 1	3
Product 2	Product 2	4
Product 5	Product 3	5

Figure 6.2 – Recommendations from the batch and light online model

We can now compute the **Precision@k** metric for the user from the two recommendations. For simplicity, let's assume the current relevant items for the user are [Product 10, Product 9, Product 1, Product 2, Product 3]. However, the model in production shows three of these relevant items and two non-relevant items – so, in this case, P(k=5) = 3/5 = 60%. The developer will look at the value of this metric and identify that the model performance has degraded. Therefore, based on the result of a recommendation from the production server and the lightweight model in the development server, the developer can manually decide to trigger retraining.

> Precision@k
>
> Precision@k is the ratio of the total relevant results to the total recommended results in the top k recommended items. For example, let's say a recommender system has made these recommendations: [R1, R2, R3, R4, R5]. However, among these recommendations, only R1, R3, and R4 are relevant to the user. In that case, Precision@5 is P(k=5) = 3/5. Here, we are considering the top 5 recommended products, so the value of k is 5.

Automatic periodic triggers

We can also periodically trigger the retraining using some periodic scheduler tools. For example, we can use **cron expressions** to set up `cron` jobs periodically.

Cron jobs

In Unix, `cron` is used to schedule periodic jobs using **cron expressions**. The job can run periodically at a fixed time, date, or interval. A `cron` expression that is used to schedule `cron` jobs can have five fields, as shown by the five asterisks in *Figure 6.3*.

```
#  ┌──────────────────────────── minute (0 - 59)
#  │ ┌────────────────────────── hour (0 - 23)
#  │ │ ┌──────────────────────── day of the month (1 - 31)
#  │ │ │ ┌────────────────────── month (1 - 12)
#  │ │ │ │ ┌──────────────────── day of the week (0 - 6) (Sunday to Saturday;
#  │ │ │ │ │                                   7 is also Sunday on some systems)
#  │ │ │ │ │
#  │ │ │ │ │
#  * * * * * <command to execute>
```

Figure 6.3 – Cron expressions and their meanings (code screenshot captured
from Wikipedia at https://en.wikipedia.org/wiki/Cron)

The fields are the following:

- **minute**: The first field is the minute field, which indicates the minute at which the job will run. If we keep the default value, *, then it will keep running every minute. We can replace * with a number, n, to denote that the job will run every n minutes. We can also have multiple minutes separated by commas. For example, if the minute field is *n1, n2, n3*, it means the job will be scheduled to run every *n1, n2*, and *n3* minutes. We can also set a range using a hyphen (-) between the two numbers. For example, *n1-n2* means the job will run every minute between *n1* and *n2*.

- **hour**: The second field denotes the hour at which the job will run. If we keep the default value, *, then the job will run every hour. As with minutes, we can specify a number, comma-separated values of multiple numbers, or a range of numbers.

- **day of the month**: This specifies the day of the month on which the `cron` job will run. We can specify a single day, a comma-separated list of multiple days, or a range.

- **month**: This field specifies the month the `cron` job will run. We can specify a month, a list of months using commas, or a range of months.

- **day of the week**: This last field specifies the weekday on which a job will run. We can specify a particular weekday, a list of weekdays by separating the numbers with commas, or a range of weekdays.

Now, we can do some exercises to get some practice with `cron` expressions. You can learn more about `cron` expressions by visiting `https://crontab.cronhub.io/`.

Let's try some small exercises.

Exercise 6.1

Write a `cron` *expression that can be used to schedule a job that will run at 6.00 AM every Saturday.*

Solution: The job will run every Saturday, so we have to set the weekday to 6; the job will run at 6 AM, so we have to set the hour field to 6; and the job will run at 0 minutes, so we have to set the minute field to 0. The expression will be 0 6 * * 6. We can verify this by going to `https://crontab.cronhub.io/` and pasting the expression, and we will see the output shown in *Figure 6.4*.

```
0 6 * * 6                At 06:00 AM, only on Saturday
```

Figure 6.4 – The semantic meaning of the cron expression 0 6 * * 6 from https://crontab.cronhub.io/

Exercise 6.2

Write a `cron` *expression that can be used to schedule a job to run at 10.30 AM on day 1 of the months of March, June, September, and December.*

Solution: The minute field will be 30 and the hour field will be 10 to run at 10.30 AM. The job should run only on day 1 of the month, so we have to set the third field to 1. The job will only run in the months of March (3), June (6), September (9), and December (12), so we need a comma-separated list of the months in the month field. The `cron` expression for this will be 30 10 1 3,6,9,12 *. We can verify this at `https://crontab.cronhub.io/` as shown in *Figure 6.5*.

```
30 10 1 3,6,9,12 *       At 10:30 AM, on day 1 of the month,
                         only in March, June, September, and
                         December
```

Figure 6.5 – The semantic meaning of the cron expression 30 10
1 3,6,9,12 * from https://crontab.cronhub.io/

To schedule a `cron` job, you can manually set the job from your terminal using the `crontab -e` command. Then, you can enter your command along with the `cron` expression.

You can see the list of your `cron` expressions using the `crontab -l` command in the terminal. The command and the response are as follows:

```
$ crontab -l


* * * * * echo Hello World! | wall
```

This command runs a job that prints the message "`Hello World!`" in the terminal every minute. The message is printed as follows:

```
Broadcast Message from mislam@Johiruls-Air.lan
        (no tty) at 22:32 CDT...
Hello World!
```

To schedule `cron` jobs with Python, you can use a library called `python-crontab`. You need to install the library using the `pip3 install python-crontab` command. To schedule the same job as before, we can now use this library:

```
from crontab import CronTab


cron = CronTab(user=True)
job = cron.new(command='echo Hello World! | wall')
job.minute.every(1)
# cron.remove_all() ## Command to remove all the scheduled jobs
cron.write() # Write the cron to crontab
```

In this code, we set a job to echo a message in the terminal every minute. The `job.minute.every(1)` statement translates to a `cron` expression of `* * * * *`. The `cron.remove_all()` command is used to remove all the `cron` jobs. To learn more about the `python-crontab` library, please visit `https://pypi.org/project/python-crontab`.

In this way, we can schedule periodic training during batch model serving. The advantages of this approach are the following:

- Training is triggered at a certain interval automatically, so no manual effort is needed and the developer team does not have to deal with the headache of observing the model

- An automatic trigger can be created easily using a simple `cron` expression

- We can tell the business users or the clients when the model will have new updates with certainty

The disadvantages of this automatic trigger are the following:

- The update may occur even though it is not needed yet, causing unnecessary use of resources.

- It might stay unnoticed if the `cron` job has failed and it may be hard to debug. However, looking into the saved logs for the given `cron` job can provide some hints for debugging. You can read about the reasons for the failures of `cron` jobs here: `https://blog.cronitor.io/4-common-reasons-cron-jobs-fail-ce067948432b`.

Using continuous model evaluation to retrain

We can also trigger the update of the model using the continuous evaluation pattern that we discussed in *Chapter 4, Continuous Model Evaluation*. We can set up a continuous evaluation component and trigger the batch update if the metric under evaluation goes below a certain threshold. For example, let's say the continuous model evaluation is set up to monitor a recommendation system. The metric being monitored is `Precision(k=5)`. The continuous evaluator will store the recommendation for some random customers, collect a survey for those users, and keep monitoring the decrease in `Precision(k=5)` performance in a dashboard. If the score is below a pre-selected threshold, then the training will be triggered as one of the actions taken by the continuous evaluator.

This trigger can happen at different times without following a fixed periodic pattern. The advantages of this approach are as follows:

- The batch update of the model takes place only when needed. In this way, we can save some computational overhead.

- The end-to-end process of updating the model is automated using a continuous monitoring system. This reduces manual effort on our part.

The disadvantage of this approach is that we need to integrate a separate continuous monitoring block, but in batch processing, since we do not expect the model to update frequently, continuous monitoring may require a lot of manual effort to collect user feedback continuously.

Based on the batch data usage strategy, batch serving can be divided into the following categories:

- **Use the full data available at the time of update**: In this case, we use all the data available up to this point to retrain the model. We consider all the data to be valuable, and with more and more data, the model becomes more robust.

- **Use a partial amount of data**: In this case, we do not use the full amount of data. We might use data from a certain time window, such as the last 2 months. Old data gets stale most of the time and does not reflect the recent situation. For example, the price of houses may become stale after a certain period.

Based on the inference strategy used, the batch serving can be divided into two different categories:

- Serving for offline inference
- Serving for on-demand inference

Let's see them both.

Serving for offline inference

During the training time for offline inference, we made inferences for different customers and stored them in a database, so whenever the inference is used, the client gets the inferences from the database instead of from the model directly. Usually, this is done in the following cases:

- **Product recommendations**: Recommendations for all customers are made during training for batch serving and stored in the database. The application that uses this recommendation makes API calls to get the records from the database. The client application does not directly interact with the model. In this case, the model even doesn't have to expose any endpoints to the clients.

- **Computing the sentiment score**: We compute the sentiment score of different articles offline. Then, the scores for those articles are stored in a database to be fetched later by client applications.

Serving for on-demand inference

In this case, although we update the model with new data periodically, we only make inferences when they are needed. For example, let's say we're predicting the weather. We might use new data to update the model to make it more intelligent, but we might be predicting on demand. Whenever a customer asks for a prediction, we send the feature to the model to get the prediction. In this case, the model needs to expose APIs to the clients.

In this section, we have discussed different categories of batch serving based on scheduling, data usage, and inference strategy. We have also introduced `cron` expressions to schedule batch jobs.

Example scenarios of batch model serving

So far, we have had a look at different categories of batch serving. In this section, we will discuss some examples of scenarios where batch serving is needed. These scenarios cannot be satisfied by online serving for the following reasons:

- The data is too large to update the model online
- The number of clients that need the inference from the model at the same time is large

For these main reasons, we will fail to satisfy both the latency and throughput requirements for the customers. Therefore, we have to do batch serving. Two example scenarios are described here.

Case 1 – recommendation

Product recommendations for clients, advertisement recommendations for users, movie recommendations, and so on are examples of where batch serving is needed. In these cases, inferences are made offline periodically and the inference scores are stored in a persistent database. The recommendations are provided to the clients from this database whenever they are needed. For example, imagine we have an e-commerce application making product recommendations to its users. The application will pull the recommendations for each of the users from a database and the recommendations in the database will be updated periodically. The model collects information about the users' purchase histories and the products' features, and at the time of the periodic updates, the model is updated and the recommendation scores for the users are computed offline. If we want to compute this recommendation online, we will fail, because there are millions of products and millions of users, which means we cannot provide the recommendations within the small latency requirement.

Case 2 – sentiment analysis

When we conduct sentiment analysis on a topic, we have to do the analysis offline, as the quantity of data may be very large. The quantity of textual data is growing very quickly due to the wide use of social media, blogs, websites, and so on. Therefore, we have to collect this data and store it to train the model periodically. The sentiment analysis response can be stored in a database and then can be retrieved by the client application quickly.

Now that we have discussed some cases where batch serving is needed, we will discuss the techniques and steps in batch model serving along with an end-to-end example.

Techniques in batch model serving

In batch serving, the main steps are as follows:

1. Set up a periodic batch update of the model.

2. Store the predictions in a persistent store.

3. The web server will pull the predictions from the database.

In this section, we will go through a batch update model serving example. We will have a file, `model.py`, that will write some random scores for five dummy products for a hypothetical customer to a CSV file. We will set up a `cron` job that will run this `model.py` file every minute and we will fetch the customer's data from a web server created using Flask.

Setting up a periodic batch update

We have already discussed that we can set up a periodic job using `cron` expressions. Usually, within a `cron` expression, we will run a script that will fetch the data from a database, then train a model, and then after doing the inference, will write the inference to a database:

1. As a demo of scheduling a batch job using `cron` expressions, let's create a Python script with the following code:

```python
import pandas as pd
import random


random_scores = []
for I in range(0, 5):
    x = round(random.random(), 2)
    random_scores.append(x)


df = pd.DataFrame(
    {
        ""product"": ""Product "",""Product "",""Product "",""Product "",""Product ""],
        ""score"": random_scores
    }
)
df.to_csv""predictions/predictions.cs"")
```

This program prints some random scores for a customer for five dummy products to a CSV file. In this program, we are creating pandas DataFrame with random scores for the following products: `["Product A", "Product B", "Product C", "Product D", "Product E"]`. The DataFrame is written to a CSV file in the `df.to_csv""predictions/predictions.cs"")` line. The program's directory structure is shown in *Figure 6.6*. The CSV file is present inside the `predictions` folder:

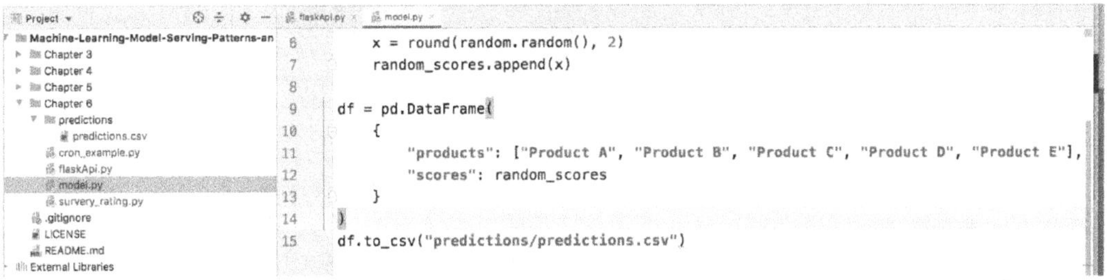

Figure 6.6 – Directory structure of model.py

2. Now, we have to run this model periodically. In order to observe change quickly, we want to run the file every minute. As we have learned, to run a job every minute, the `cron` expression will be `* * * * *`. To enter the `cron` expression, we need to go to the terminal and then type in the `crontab -e` command, as shown in *Figure 6.7*.

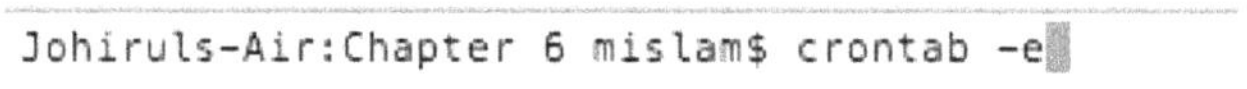

```
Johiruls-Air:Chapter 6 mislam$ crontab -e
```

Figure 6.7 – To edit crontab, the crontab –e command needs to be used

3. After the command, we will see an editor window in which we can enter our command. Enter the command as shown in *Figure 6.8*.

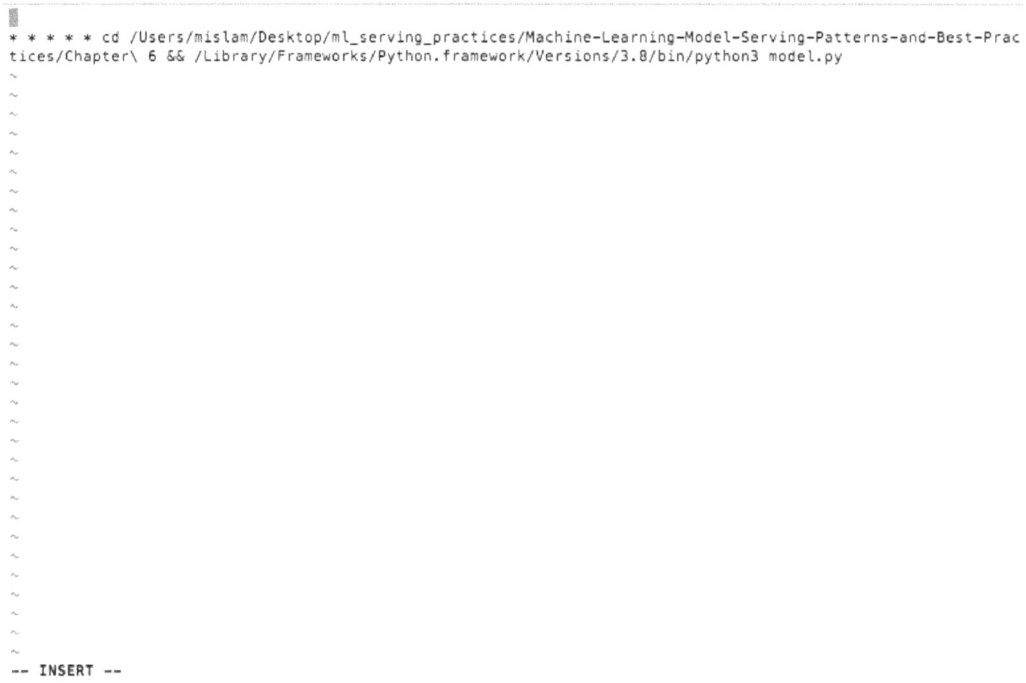

```
* * * * * cd /Users/mislam/Desktop/ml_serving_practices/Machine-Learning-Model-Serving-Patterns-and-Best-Prac
tices/Chapter\ 6 && /Library/Frameworks/Python.framework/Versions/3.8/bin/python3 model.py
~
~
~
~
~
~
~
~
~
~
~
~
~
~
~
~
~
~
~
~
~
~
~
-- INSERT --
```

Figure 6.8 – The cron expression added to the crontab for running model.py

The expression is as follows:

```
* * * * * cd /Users/mislam/Desktop/ml_serving_practices/
Machine-Learning-Model-Serving-Patterns-and-Best-
Practices/Chapter\ 6 && /Library/Frameworks/Python.
framework/Versions/3.8/bin/python3 model.py
```

Here, the * * * * * part indicates that the following command will run every minute. Here, we are using two commands. The first command, cd /Users/mislam/Desktop/ml_serving_practices/Machine-Learning-Model-Serving-Patterns-and-Best-Practices/Chapter\ 6, is used for moving to the Chapter 6 folder, where my Python file is present. The second command, /Library/Frameworks/Python. framework/Versions/3.8/bin/python3 model.py, is used to run the model. py file. Notice that we are using the full path of the python3 binary to run the Python file. You can find the full path to the python3 binary using the which python3 command.

Now, we have scheduled our periodic batch job using a cron expression. Our dummy job will run every minute, but you may schedule jobs with a larger periodic interval.

Storing the predictions in a persistent store

In our model.py program, we are storing the dummy predictions inside a CSV file. If we have a large number of customers, it is ideal to store the predictions in the database.

Let's check whether our predictions are being written to the CSV file at the expected interval scheduled using the cron job.

If we go inside the predictions folder, we will notice that a CSV file called predictions.csv is being written to every minute.

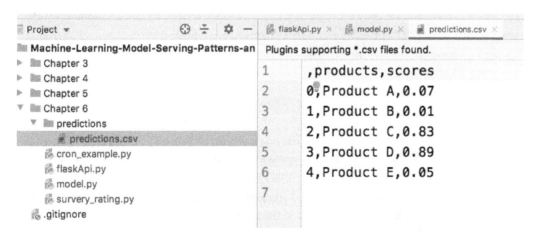

Figure 6.9 – The CSV file being written to periodically by the batch job

The file and directory structure of the file are shown in *Figure 6.9*.

Pulling predictions by the server application

When serving the application that uses inferences to the client, the server gets the inferences from the persistence store. The server does not make inference requests directly to the model:

1. To show the server application fetching recommendations from the database, let's create a basic server using the Flask API as follows:

```
import pandas as pd
from flask import Flask, jsonify

app = Flask(__name__)

@app.route""/predict_batc"", methods=''POS''])
def predict_loading_params():
    df = pd.read_csv""predictions/predictions.cs"")
    predictions = []
    for index, row in df.iterrows():
        product = row''product'']
        score = row''score'']
        predictions.append((product, score))
    predictions.sort(key = lambda a : a[1])
    return jsonify(predictions)

app.run()
```

You can see that the server application has one endpoint, "/predict_batch". Once an application hits this endpoint, the server opens a connection to the persistent store. In our case, we open the CSV file to which we are writing predictions in batches.

If we run this application by running the Python program that contains the app.run() code snippet, we will see the following output in the terminal:

```
* Serving Flask app "flaskApi" (lazy loading)
* Environment: production
  WARNING: This is a development server. Do not use it
in a production deployment.
  Use a production WSGI server instead.
* Debug mode: off
```

```
* Running on http://127.0.0.1:5000/ (Press CTRL+C to
quit)
127.0.0.1 - - [23/Jun/2022 21:41:52] "POST /predict_batch
HTTP/1.1" 200 -
```

2. Now, we can go to Postman to test the endpoint. The Postman interface, after calling this endpoint, will look like *Figure 6.10*.

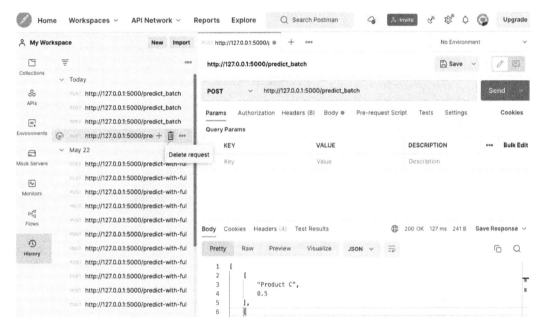

Figure 6.10 – Using Postman to call the http://127.0.0.1:5000/predict_batch API

We have scheduled the batch job to update the CSV file every minute with random scores. The output we get in the first run is the following:

```
[
["Product C", 0.5 ],
["Product E",0.82],
["Product A",0.89],
["Product B",0.94],
["Product D",0.97]
]
```

The responses are sorted in increasing order of score.

3. If we make the call again after some time, we will get new predictions. After 2 minutes, we make another call and get the following result:

```
[
["Product B",0.06],
["Product E",0.15],
["Product D",0.39],
["Product C",0.58],
["Product A",0.7]
]
```

The scores and the ranking of the recommendations have also changed.

4. Ideally, the scores will be sorted in reverse order and the recommendation with the highest score will be at the top. To do that, you can add `reverse = True` to `model.py`, as follows:

```
predictions.sort(key = lambda a : a[1], reverse = True)
```

The output will now be printed in reverse order. The product with the highest score will come at the beginning of the list as follows:

```
[
["Product A",0.7],
["Product C",0.58],
["Product D",0.39],
["Product E",0.15],
["Product B",0.06]
]
```

In this section, we have talked about the techniques in batch model serving by showing an end-to-end dummy example. In this next section, we will discuss some limitations of the batch model serving.

Limitations of batch serving

Batch serving is essential in today's world of big data. However, it has the following limitations:

*   *Scheduling the jobs is hard*: Scheduling periodic batch jobs is sometimes complicated. As we have seen, during scheduling, the paths expected by the `cron` expression need to be given carefully. Mostly, `cron` expressions expect absolute paths. The scheduled jobs may also introduce a single point of failure. If somehow it fails to run on schedule, we might not have the latest inferences, causing a bad customer experience.

- *Growth of data will make training slow*: If the data grows, the training may gradually take more time. For example, the time needed to train a model with 10 MB of data will not be the same as the time needed to train a model with 10 GB of data. Therefore, we need to take care of this scenario. In most cases, we can discard old data, as it will become stale. Then, the question arises, how old is the data when we consider it stale? We need to identify the answer through empirical evaluation. In some cases, we can say that 1-month-old data is stale and should not be used; in other cases, it can be multiple years.

- *Cold start problem*: When a new client starts using an application, the user will not have any history that can be used to make recommendations for the user. For example, a new user starts using an e-commerce site. The site will not be able to provide recommendations for the user, as there is no history available for the user and the model will not update until the next scheduled time of training comes. To solve this problem, there are some approaches to provide an initial recommendation. For example, we can look at the user's age group and other information provided during sign-up to make recommendations from other users with similar personal information.

- *Need to maintain a separate database for predictions*: We have to store the inference in a separate table or database during batch predictions. This creates additional maintenance overhead. Furthermore, to distinguish between separate batches of inferences, we need to use another column in the prediction table, indicating the batch. Sometimes, we might need to maintain separate tables for storing separate batches of predictions. All these can give rise to technical challenges and technical debts if not handled carefully.

- *Monitoring*: Monitoring batch serving is also a challenge. As the process runs for a long time, the window of failure is longer. It is also a technical challenge to separate the run between triggered and scheduled batch jobs.

In this section, we have seen some limitations of batch model serving. In the next section, we will draw the summary of the chapter and conclude.

Summary

In this chapter, we have talked about the batch model serving pattern. This is the first pattern that we have categorized based on the serving approaches we discussed in *Chapter 2, Introducing Model Serving Patterns*. We introduced what batch serving is and looked at the different batch serving patterns based on the batch job triggering strategy and the inference strategy. We explained, with examples, how `cron` expressions can be used to schedule batch jobs. We discussed some examples where batch serving is essential. Then, we looked at an end-to-end example of batch serving using a dummy model, persistence store, and server.

In the next chapter, we will discuss online serving, which updates the models immediately with new data.

Further reading

We can use the following reading materials to further enhance our understanding:

- To find out more about different metrics in recommender systems, visit `https://neptune.ai/blog/recommender-systems-metrics`

- To learn more about `cron` expressions and scheduling `cron` jobs, visit `https://ostechnix.com/a-beginners-guide-to-cron-jobs/`

Online Learning Model Serving

In online model serving, we want the impact on the model of recent data to be visible as soon as the data is available. In some cases, during online training, retraining of the model may happen during the serving of each response. This is done to ensure that the model is up to date with newly arrived data. In this chapter, we will talk about the online model serving pattern. We will introduce you to online model serving, the advantages and disadvantages of online model serving, and techniques for online model serving.

In this chapter, we will cover the following topics:

- Introducing online model serving
- Use cases for online model serving
- Challenges in online model serving
- Implementing online model serving

Technical requirements

In this chapter, we will mostly use the same libraries that we have used in the previous chapters. You should have Postman or another REST API client installed to be able to send API calls and see the response. All the code for this chapter is provided at this link: `https://github.com/PacktPublishing/Machine-Learning-Model-Serving-Patterns-and-Best-Practices/tree/main/Chapter%207`.

If you get `ModuleNotFoundError` while trying to import any library, then you should install that module using the `pip3 install <module_name>` command.

Introducing online model serving

In online model serving, the model is updated automatically along with the user input as a backend process of retraining the model. So, the response from the model reflects the most recent data available for training. Whenever we send a prediction request to the model with an input, the model updates the model weights and biases by running the training for the provided input, and at the same time, the prediction response is provided.

We can have a look at *Figure 7.1* to see how the model update is coupled with the prediction requests in the online model serving. Whenever a prediction request is made, it also involves updating the model:

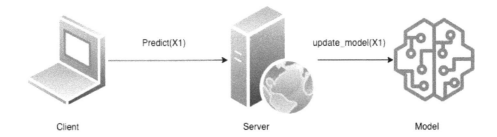

Figure 7.1 – Online model serving usually couples the model updating with the prediction requests

Let's go into more detail on serving.

Serving requests

We have seen that in online model serving, each prediction request also triggers an update model request. Now, this question comes to mind: when does the model send the response to the request? The serving of the request can be divided into two categories:

- Sending the prediction response first and then updating the model
- Updating the model first and then sending the response

We will discuss these two modes of serving prediction requests in the following paragraphs.

Sending the prediction first

In this approach, when a client sends a request for prediction with the input, the server first returns the response using the last trained model and then does the update. Therefore, the client gets the response from the old model. This is ideal practice, as it preserves the unseen nature of the input during prediction.

The advantages of this approach are the following:

- The client gets a quicker response as the response does not need to wait for the update of the model.

- It preserves the unseen nature of the prediction data. This way, we can avoid overfitting the model with all the data. For example, if we retrain the model with the input the user has passed for prediction, the model is supposed to memorize the new data and provide a biased response.

- This approach can work in both supervised and unsupervised learning.

For these reasons, this is the most desirable approach if the prediction request contains only a few inputs (most of the time, there may be just one input) during a request.

However, after sending the response, the model gets updated for every subsequent request.

To demonstrate serving the request and then updating the model, let's take a simple model that can be found here: `https://keras.io/examples/vision/mnist_convnet/`

1. To make the model simple, we modify it as follows:

```
model = keras.Sequential(
    [
        keras.Input(shape=input_shape),
        layers.Conv2D(32, kernel_size=(3, 3),
activation="relu"),
        layers.MaxPooling2D(pool_size=(2, 2)),
        layers.Flatten(),
        layers.Dense(num_classes, activation="softmax"),
    ]
)
```

2. For training the model, we also reduce the data size by taking only the first few samples using the following code:

```
x_train = x_train.astype("float32") / 255
x_test = x_test.astype("float32") / 255
# Make sure images have shape (28, 28, 1)
x_train = np.expand_dims(x_train[0:1000], -1)
x_test = np.expand_dims(x_test[0:100], -1)
print("x_train shape:", x_train.shape)
print(x_train.shape[0], "train samples")
print(x_test.shape[0], "test samples")
```

```
y_train = keras.utils.to_categorical(y_train[0:1000],
num_classes)
y_test = keras.utils.to_categorical(y_test[0:100], num_
classes)
```

We take only 1,000 samples for training and 100 samples for testing.

3. In this code, we just kept four layers for simplicity. We save the trained model using this code:

```
model.save("keras_mnist", save_format="tf")
```

4. We will see that after the training, the model is saved along with all assets to a folder named keras_mnist, as shown in *Figure 7.2*:

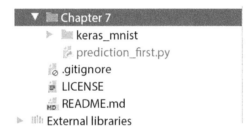

Figure 7.2 – The trained model is saved for online serving

The model we have trained is not a good model and is highly underfitted. Its performance is shown in the output as follows:

```
Test loss: 1.923134207725525
Test accuracy: 0.5899999737739563
```

As we want to demo updating the model live as new data comes in, it is fine for us.

Now let's load the saved model and assume it is being loaded inside a server. To demonstrate how the model is updated after prediction, we take the following steps:

5. We load the saved model before using the following code snippet:

```
import keras
import numpy as np

model: keras.models.Sequential = keras.models.load_
model("keras_mnist")
print(model.summary())
```

We will be able to see the following model summary from the last `print` statement:

```
Layer (type)                 Output
Shape                Param #
===========================================================
conv2d (Conv2D)              (None, 26, 26,
32)          320


max_pooling2d (MaxPooling2D) (None, 13, 13, 32)          0


flatten (Flatten)            (None, 5408)                0


dense (Dense)                (None,
10)              54090
===========================================================
Total params: 54,410
Trainable params: 54,410
Non-trainable params: 0
```

Therefore, the model is loaded correctly.

6. Let's use the first 10 samples from the `mnist` test dataset to make predictions online. In this step, we prepare the data using the following code:

```
x_test = x_test.astype("float32") / 255
x_test = x_test[0:10]
y_test = y_test[0:10]
y_test = keras.utils.to_categorical(y_test, 10)
# Make sure images have shape (28, 28, 1)
x_test = np.expand_dims(x_test, -1)
```

7. Make sure the formatted data is in the correct shape by using the following snippet:

```
print(x_test.shape)
print(y_test.shape)
```

You will get the following output from the preceding two lines:

```
(10, 28, 28, 1)
(10, 10)
```

8. Let's compute the prediction score from this model using the following code snippet:

```
score = model.evaluate(x_test, y_test, verbose=1)
print("Test loss:", score[0])
print("Test accuracy:", score[1])
```

As we know the correct labels for x_test dataset, we are able to compute the accuracy of prediction here. We will see the following output and notice that the prediction accuracy is very low as we are using an underfitted model:

Test loss: 1.8864532709121704

Test accuracy: 0.800000011920929

9. Now, we have to update the model with the data that we had. To update the model, we can use the following code snippet:

```
model.fit(x_test, y_test, batch_size=100, epochs=15,
validation_split=0.1)
score = model.evaluate(x_test, y_test, verbose=0)
print("Test loss:", score[0])
print("Test accuracy:", score[1])
```

We used x_test and y_test to train the model. We improved the accuracy a little after this update, and we now see the following output from the last two print statements:

Test loss: 0.3599579930305481

Test accuracy: 0.8999999761581421

> **Note**
>
> The evaluation result shown after updating the model in *step 4* is biased as we used the same data for training and testing. We used this just to show that the model was updated from its last state.

10. Now, to ensure that the model actually was updated, we can save the updated model to a different folder using the following code snippet:

```
model.save("keras_mnist2", save_format="tf")
```

11. Now, let's load both models using the following code snippet and check the difference in the weights of both models:

```
model1: keras.models.Sequential = keras.models.load_
model("keras_mnist")
W1 = model1.get_weights()
```

```
model2: keras.models.Sequential = keras.models.load_
model("keras_mnist2")
W2 = model2.get_weights()
diff = W2[0] - W1[0]

print("Printing diff")
print(diff[0][0][0])
```

For simplicity, we only printed a small portion of the weight difference using the code, and we got the following output:

```
Printing diff
[ 0.01194626   0.01476786   0.01497105   0.01482136
0.00947043   0.01504258

   0.01488613   0.01489267   0.01124015   0.01478248
0.01488066   0.00905368

   0.01482182   0.01473019  -0.00484286   0.00875221
0.01496141   0.01464234

   0.01482094   0.0146517    0.01486789   0.014983
0.01485247   0.0148745

  -0.01209575   0.01469799   0.0147083    0.01447299
0.01497509   0.01459881

   0.01506994   0.01482368]
```

So, we should notice that there is a significant difference in weights between the two versions of the model. We should save the new version of the model to the same location from where the model is served to ensure that, in subsequent requests, the new model is used.

You'll notice that while updating the model, we used x_test and y_test. However, in a real-life scenario, we might not know the actual label for the input data. Then how can we update the model? We can update the model in this case in one of the following ways:

- Ask for feedback from the user after the prediction result is received. If the feedback is in favor of the prediction result, then you can use that prediction to update the model. For example, let's say that we have a flower classification problem that classifies either *rose* or *jasmine*. Now, we get an inference request from the client with input instance of [X1, X2, X3, X4, X5]. To use these new inputs for training, we need labels for them. We can get these labels via predictions and ask for customer feedback if the results are correct. If we provide the predictions [Rose, Jasmine, Rose, Rose, Jasmine] and the customer provides positive feedback on the prediction result, then we can use the labeled data [(X1, Rose), (X2, Jasmine), (X3, Rose), (X4, Rose), (X5, Jasmine)] to update the model. The feedback can be given in the form of a rating out of 10 by the user. We can set a criterion that we will only use the data to update the model if the feedback from the user is greater than 9. Otherwise, we will discard the data. This kind of idea can be used for recommendations, online translations, and so on. We can ask for customer feedback and based on the feedback, we can update the model if needed.

- Collect new data with labels using manual effort and update the model once the labels have been collected. After collecting the labeled data, we can store it in a location, and the model can load the data from that location and train. In this case, after every prediction, instead of using the input data directly for training, the model can look for new data in the directory. If there is new data, only then can the model start updating. After the data has been used, the training program can move the file from that location to some other location or delete the file if it is okay with the team. This is done so that the model does not check the same data again and again by mistake. We could have a dummy code snippet like the following where we write the file, and when a model update is requested, we first check whether new data is available:

```python
import os

def update_model():
    if os.path.exists("train_directory/data.csv"):
        data = open("train_directory/data.csv", "r")
        print("Training the model with new data")
        os.remove("train_directory/data.csv")
        print("File deleted after training")
    else:
        print("No new data to train")

def create_file():
    file = open("train_directory/data.csv", "w")
    file.write("X, Y\n")
    file.write("100, 1\n")
    file.write("200, 2\n")


if __name__ == "__main__":
    # create_file() # Uncomment this to create the file
    update_model()
```

After the training is done, we delete the file. If we look into the code snippet in the `update_model()` function, we first check whether the file exists in the `if` block. We start training if the file exists, and after the training is completed, we delete the file. There is another function, `create_file()`, that is used to create the file. This function can be used by some other process to write data. In our code, we have shown it in the same file for understanding. We call both functions from the main block. We commented out the `create_file()` function call. We can uncomment it if we want to create a file. If the file is present, we will see the following output from the `if` block:

```
Training the model with new data
File deleted after training
```

And if the file is not present, we will see the following output:

```
No new data to train
```

So far, we have seen how we can update the model after training, along with some basic examples to demonstrate the workflow step by step. However, other than the data labeling issues, we have the following challenges that need to be handled while updating:

- A model needs to serve multiple requests at the same time. If we want to start multiple training at the same time, then we might break the system. To solve this challenge, we can store the prediction requests and, for a number of requests, do a single update. And until that update is finished, we can lock further update requests.

- A model update may make the model worse if the new data is not good. To tackle this issue, we can add some data validation techniques before updating the model.

In this section, we have seen the technique of updating the model after the prediction is sent to the user from the old model. In the following sub-section, we will see the technique of updating the model first and then using the updated model for sending predictions. This approach can be used in unsupervised models, as we do not have labeled data at the time of prediction to train a supervised model.

Updating the model first

In this approach, usually, the request comes with an array of inputs, such as [X1, X2, X3, X4, ..., Xn]. We can use a part of the randomly selected data to retrain the model and then send responses for all of them as [Y1, Y2, Y3, Y4, ..., Yn]. We can't train a supervised model in this case. The model needs to be unsupervised to update the model before prediction because we will not be able to know the labels of the input before the prediction is made. However, there is still some chance of overfitting for this case for the data points that are used during training. To demonstrate how the unsupervised model is updated first before sending a response, take the following steps:

1. Build a dummy KMeans clustering model as follows:

```
from sklearn.cluster import MiniBatchKMeans
import numpy as np
```

```
X = np.array([[1, 2], [1, 4], [1, 0],
              [4, 2], [4, 0], [4, 4],
              [4, 5], [0, 1], [2, 2],
              [3, 2], [5, 5], [1, -1]])


kmeans = MiniBatchKMeans(n_clusters=2,
                         random_state=0,
                         batch_size=6).fit(X)
```

2. To update the model online we use the following code snippet:

```
def update_model(X):
    global kmeans
    kmeans = kmeans.partial_fit(X)
    print("New Cluster centers are")
    print(kmeans.cluster_centers_)
```

3. The following function is used to make a prediction from the model. Before making a prediction, we update the model with the new data in this example:

```
def predict(X):
    # Train the model first
    update_model(X)
    predictions = kmeans.predict(X)
    print(predictions)
```

4. Then, we call the `predict` function with four different sets of dummy data in the following code snippet:

```
if __name__ == "__main__":
    predict([[0, 0], [4, 4]])
    predict([[1, 0], [1, 4]])
    predict([[0, 1], [2, 4]])
    predict([[0, 3], [3, 4]])
```

The example is collected and modified from this link https://scikit-learn.org/stable/modules/generated/sklearn.cluster.MiniBatchKMeans.html.

In this code, whenever we send a request to make a prediction, the KMeans model first gets updated with the input data. We made four prediction requests in the main block in the preceding program, and we got the following output:

```
New Cluster centers are
[[3.96        2.44       ]
 [1.0952381  1.35714286]]
[1 0]
New Cluster centers are
[[3.96        2.44       ]
 [1.09090909 1.38636364]]
[1 1]
New Cluster centers are
[[3.92156863 2.47058824]
 [1.06666667 1.37777778]]
[1 0]
New Cluster centers are
[[3.90384615 2.5        ]
 [1.04347826 1.41304348]]
[1 0]
```

Notice that the cluster centers of the KMeans model are updated after every prediction request. In this model, we only used two clusters. We notice in the second request the [3.96 2.44] cluster center for the first cluster is still the same as from the first request. So, the update was not necessary for the first cluster for the incoming prediction data [[1, 0], [1, 4]].

This kind of pattern can be used in critical systems such as autonomous driving, which can learn about new patterns before sending inferences for an unseen pattern to enhance security and other sensitive unsupervised systems. The drawbacks of this approach are as follows:

- It can only be used effectively in unsupervised models.
- The update part of the model can add to the latency of the serving and cannot be used for updating a large number of inputs and complicated models.
- The response from the model will be highly biased as the same data passed for prediction is used to train the model first.

In this section, we have introduced online model serving to you along with some techniques for updating the model using the online serving pattern. We have also discussed that the updating of the model might also take some manual data labeling efforts and data validation strategies to ensure a successful update. In the next section, we will discuss some example cases where online model serving is more suitable than batch serving.

Use cases for online model serving

Online model serving is essential when we need to see the impact of recent data on the model as soon as possible instead of waiting for a periodic batch update. In this section, we will discuss the following example cases where online serving is needed:

- Recommending the nearest emergency center during a pandemic
- Predicting the favorite soccer team in a tournament
- Predicting the path of a hurricane or storm
- Predicting the estimated delivery time of delivery trucks

Case 1 – recommending the nearest emergency center during a pandemic

Let's assume a lot of people are becoming sick due to a pandemic, such as Covid, every day. However, there is a limited number of hospitals. We want to recommend the nearest, most convenient hospital to a patient, keeping the following things in mind:

- The distance between the patient and the hospital should be short
- The number of patients sent to a particular hospital needs to be within a certain capacity

We can train a k-means clustering model where the cluster centers will denote which hospital the patient should go to. All the patients in a cluster will be recommended to get treatment from the hospital that is the closest to the center of the cluster. As the number of patients is regularly increasing, we need to update the model every time a new patient tests positive for the disease. The model will be updated with features, such as the address, the geolocation of the patient, and so on. After a prediction response is sent regarding which hospital this patient should go to, the model should also be updated. To demonstrate this case, let's modify the preceding k-means code a little in the following way:

```
X = np.array([[0, 0], [0, 1]])

kmeans = MiniBatchKMeans(n_clusters=2,
                         random_state=0,
                         batch_size=6).fit(X)
print("Initial cluster centers")
print(kmeans.cluster_centers_)

# Same update and predict functions as before are ommitted
if __name__ == "__main__":
    predict([[0, 0], [0, 1]]) # Ra
```

```
predict([[10, 10], [10, 15]])
predict([[10, 11], [11, 15]])
predict([[11, 10], [11, 14]])
predict([[0, 0], [0, 1]]) # Rb
```

We notice the output from the two `predict` requests marked as Ra and Rb in the preceding code snippet are the following:

New cluster centers are shown as follows:

```
[[0. 1.]
 [0. 0.]]
[1 0]
```

New cluster centers are as follows:

```
[[1.46511628 2.60465116]
 [0.         0.02857143]]
[1 1]
```

Look at the highlighted parts in the outputs. So, the [0, 1] data point has changed its cluster now. If we did not do the update online, the model would still have predicted the 0 cluster for the [0, 1] data point.

A patient's recommended hospital may change pretty quickly as the number of patients is increasing a lot due to the pandemic. Therefore, we need to update the model online.

Case 2 – predicting the favorite soccer team in a tournament

Let's imagine that the FIFA World Cup tournament is going on. A lot of people are interested to know which team is the favorite to win the final match, based on the ongoing performances of all the teams in different individual matches. The model that is used to make this prediction needs to update after every single match. So, if a client sends a request for the prediction of the favorite team after a match, the model should update itself after that.

In this way, training of the model using newly available data is done immediately after the prediction request is made. The model could be a recommendation model that returns the top k recommendations.

For example, let's consider a simple model that returns the favorite team in terms of the average number of times the team kept control of the soccer ball during a match. Let's say there are three teams – *Team1*, *Team2*, and *Team3*. The data so far is shown in the following table, *Figure 7.3*:

Team	#Matches	#TotalShotsAttempted	#AvgShotsAttempted
Team1	3	30	10
Team2	2	24	12
Team3	2	18	9

Figure 7.3 – Information on ball control time of different teams in the tournament

We notice that the most favorable team is **Team2** as per the data provided in the table.

> **Note**
> **#TotalShotsAttempted** in *Figure 7.3* denotes the total number of shots on goal the team attempted over the matches it has played. **#AvgShotsAttempted** denotes the average number of shots the team made for goals over the number of matches they have played.

Team	#Matches	#TotalShotsAttempted	#AvgShotsAttempted
Team1	3	30	10
Team2	3	27	9
Team3	3	36	12

Figure 7.4 – Information on ball control time of different teams in the tournament

Let's assume just now a match between **Team2** and **Team3** is completed and the new data is shown in *Figure 7.4*. The **#Matches** for both **Team2** and **Team3** is now **3**. The total number of shots attempted by **Team3** is **18** in the third match and the total shots attempted by **Team2** is **3** in the third match. So, if we ask the model to predict the favorite team and the model is not updated with the data of the third match and still uses the old data as shown in *Figure 7.3*, the model will make the wrong prediction for **Team2** using the old data, because the model still thinks **#AvgShotsAttempted** for **Team2** is **12** and it is the most favorable team. However, in reality, after the third match #TotalShots for Team2 is **27** in 3 matches and #TotalShots for Team3 is **36** in 3 matches. So, the average for Team3 is now 36/3 ~ 12, the average for Team2 is 27/3 ~ 9, and the average for Team1 is still **10**. These new statistics are shown in the table in *Figure 7.4*. So, in reality, the favorite team now is Team3, which was the least favorable before the last match.

That's why updating the model online in this scenario is very important.

Case 3 – predicting the path of a hurricane or storm

During natural disasters, we need to make immediate decisions. So, we might have to update the model that predicts the path of a hurricane whenever some new data is available. We can get features, such as coordinates, wind direction, wind speed, and so on, and use those in a model to predict the path and severity of the storm. If you are interested in different features that can help you to make a good machine learning model to predict the path of a live hurricane, you can follow this link: `https://arxiv.org/abs/1802.02548`.

Case 4 – predicting the estimated delivery time of delivery trucks

I guess you have seen a situation when you have ordered food to be delivered to your home, but the food was delivered later than the original estimated time. The apps that provide an estimate of deliveries frequently update the estimation based on the new data available. This is how you can stay updated on the delivery time. Similarly, in a supply chain, updating the estimated delivery time of a delivery truck can be done using online training.

In this section, we have given a high-level overview of a number of use cases where online model serving is needed. In the next section, we will discuss some of the challenges in online model serving.

Challenges in online model serving

Though online model serving is the solution to go with for a number of business cases, it also has some challenges. Some of the challenges of online model serving are described here:

- We might not be able to inject new data directly into training
- The model can start underperforming if some wrong data is passed as input
- The model may become overfitted for a particular class
- The latency might increase
- There may be concurrent update requests

We will discuss these challenges and some possible solutions to the challenges in the following sub-sections.

Challenges in using newly arrived data for training

Often, we will not be able to directly use new data to extract features and train the model. Our data might have problems, such as missing fields, wrong types, wrong dimensions, and so on. So, we have to do data wrangling or data cleaning before we can start extracting features from raw data.

For example, let's consider the following code snippet:

```python
import pandas as pd

df = pd.DataFrame({"X": [5, 6, "Seven"], "Y": [2, None, 5]})
print(df)

from sklearn.linear_model import LinearRegression
```

```
X = df['X']
y = df['Y']
model = LinearRegression().fit(X, y)
```

Here, we have "Seven" as the wrong data type in the "X" column. There is also a missing value in the "Y" column. If we tried to directly use this data in a LinearRegression model we would get the following error:

ValueError: could not convert string to float: 'Seven'

Therefore, we see that if we try to directly apply raw data to the model without data wrangling, it might fail. So, online model serving does not mean we should immediately send the raw data for training. In fact, data wrangling should also be a part of this process.

Underperforming of the model after online training

During online training, if some bad or adversarial data is inserted, then the model may start underperforming. The model may even start giving opposite predictions from the original labels.

For example, let us consider the following code snippet:

```
from sklearn.linear_model import SGDClassifier

X = [[1, 1], [1, 1], [2, 2], [2, 2]]
y = [0, 0, 1, 1]
model = SGDClassifier()
model.fit(X, y)
res = model.predict([[1, 1]])
print(res)
model.partial_fit(
    [[1, 1], [1, 1], [1, 1]],
    [1, 1, 1])
res = model.predict([[1, 1], [2, 2]])
print(res)
```

Here, we train a model to classify [1, 1] as [0] and [2, 2] as [1]. So, from the first print statement, we get the following output:

```
[0, 1]
```

Therefore, the model is able to predict [1, 1] and [2, 2] correctly.

Now let's assume the model is updated in online mode using the partial_fit function. However, we got some wrong data from an adversary. The data has labeled [1, 1] as [1] instead of [0]. Now, the model is updated with this wrong data. If we make the same prediction with the updated model, we get the following result from the last print statement:

```
[1, 1]
```

We notice the model is now failing to recognize the [1, 1] instances correctly. The model classifies them as [1] instead of [0]. This is how the model has become totally blind to one of the classes and has started underperforming.

To solve this problem, we have to do some data validation before using the new data during training. A simple verification for the problem described could be to try to do a one-pass prediction using the existing model. If there is any contradiction between the predicted value and the label of the new data, then we can send that data for further manual evaluation.

We can modify the preceding program to add a simple verification as follows (only the partial fitting/online training part of the code is shown here):

```python
def should_use_in_partial_fit(model, X, y):
    preds = model.predict(X)
    equal = np.array(y) == np.array(preds)
    size = len(equal)
    count = 0
    for x in equal:
        if x == True:
            count += 1
    avg_correct = count / size
    if(avg_correct > 0.9):
        model.partial_fit(X, y)
        return model
    else:
        print("Data label seems wrong. Please manually verify")
        print("Not updating the model with this data. Returning
the old model.")
        return model

X_new = [[1, 1], [1, 1], [1, 1]]
y_new = [1 , 1, 1]
```

```
model = should_use_in_partial_fit(model, X_new, y_new)
res = model.predict([[1, 1], [2, 2]])
print(res)
```

Here, we first check whether we should use the new data for updating the model or not. We simply do a prediction of the new data with the last model and check whether there is any predicted label that is not equal to the provided data labels for the new data. If the model fails to correctly classify at least 90% (a dummy threshold that can vary based on the problem) of the input, then we do not update the model and return the old one. Therefore, we get the following output from the preceding code:

```
[0 1]
Data label seems wrong. Please manually verify
Not updating the model with this data. Returning the old model.
[0 1]
```

The model is not updated and provides a message for the user to further verify the data.

Overfitting and class imbalance

In real life, the same kind of data may be seen more frequently. For example, let's say that, on a warm day, everyone is buying ice from a store. So, the model may be overfitted for the prediction of ice prices. Similarly, in a multi-class prediction problem, only a particular class may be coming up repeatedly as new data, making the model imbalanced to a particular class. So, in these cases, the model may be slowly biased towards a particular model.

Increasing of latency

The latency of serving a prediction request from the client might increase if the model updates before sending the response. For example, let us consider that the latency of the response without updating the model is 1 second. Now, if the online update of the model takes 2 seconds, then the updated latency for the response will be 1 + 2 = 3 seconds.

We notice that the response time can increase a lot if the training becomes expensive in terms of time. We should not train the model before sending a response if the training is expensive. And we should always start training from the last state of the model instead of retraining with all the historical data.

Handling concurrent requests

There may be many concurrent requests being made. If we attempt to update the model for each of these concurrent requests, then we might make the model corrupt. To handle this, we need to update the model after a certain number of requests are made. We can store those requests in a queue, and if the size of the queue reaches a selected threshold, we can train the model at that instant.

For example, we can use a queue to check whether a sufficient number of requests have been stored or not. We can look at the following code snippet to get an idea of how we can use a queue for training the model after only a certain number of requests:

```
import queue

Q = queue.Queue()
class Request:
    def __init__(self, id, data):
        self.id = id
        self.data = data

def update_model(request):
    Q.put(request)
    if(Q.qsize() >= 3):
        print("Now the model is going to update")
        while Q.qsize() > 0:
            req = Q.get()
        print("After training the queue size now: ", Q.qsize())
    else:
        print("Not going to update yet")

request1 = Request("id1", [[1, 1]])
update_model(request1)
request2 = Request("id2", [[1, 1]])
update_model(request2)
request3 = Request("id3", [[2, 2]])
update_model(request3)
```

Here, we add the request to a queue whenever a request is made, and we update the model once three requests are collected. We empty the queue after the training of the model is completed. From the preceding code, we get the following output:

```
Not going to update yet
Not going to update yet
Now the model is going to update
After training the queue size now:   0
```

We can see that we do not update the model for the first two requests. We update the model after the third request and clear the queue after the training. In this example, we do not call any training code. It is just for your understanding of the workflow. We can also make these methods `asynchronous` (`async`) to update them asynchronously if needed. We discussed asynchronous (async) operations in *Chapter 6*, *Batch Model Serving*.

In this section, we have covered some of the challenges in online serving and how we can mitigate those challenges. In the next section, we will discuss a dummy end-to-end example of doing online serving.

Implementing online model serving

In this section, we will train a dummy `SGDRegressor` model, use `Flask` to create a server and API for the online prediction endpoint, use Postman to send a request to the server, and update the model with the input data and the prediction made by the last model.

For the end-to-end example we are going to run, you need to import the following modules:

```
from flask import Flask, request
import numpy as np
import json
from sklearn.linear_model import SGDRegressor
```

Let's begin:

1. First of all, let's create a model with some dummy data in the following code snippet:

    ```
    X = [
        [1, 1, 1],
        [1, 1, 1],
        [1, 1, 1],
        [2, 2, 2],
        [2, 2, 2],
        [2, 2, 2]
        ]
    y = [1, 1, 1, 2, 2, 2]
    model = SGDRegressor()
    model.fit(X, y)
    print("Initial coefficients")
    print(model.coef_)
    ```

```
print("Initial intercept")
print(model.intercept_)
```

We notice that the model is trained with the X and y dummy data. We get the initial coefficients and the intercepts as the following:

```
Initial coefficients
[0.29263344 0.29263344 0.29263344]
Initial intercept
[0.16956488]
```

2. Now we create the server using Flask and a prediction API using the following code snippet:

```
app = Flask(__name__)

def update_model(Xn, yn):
    print("Updating the model")
    model.partial_fit(Xn, yn)
    print("New coefficients now")
    print(model.coef_)
    print("New intercept now")
    print(model.intercept_)

@app.route("/predict-online", methods=['POST'])
def predict_online():
    predictions = []
    X = json.loads(request.data)
    print("Input data is ", X)
    predictions = model.predict(X)
    update_model(X, predictions)
    return json.dumps(predictions, cls=NumpyEncoder)

if __name__ == "__main__":
    app.run()
```

We have the update_model function, which takes new data and updates the model using the model.partial_fit(..) function. If we run this program, we will notice in the console that the server is started, as shown in *Figure 7.5*:

```
* Serving Flask app "flaskApi" (lazy loading)
* Environment: production
  WARNING: This is a development server. Do not use it in a production deployment.
  Use a production WSGI server instead.
* Debug mode: off
* Running on http://127.0.0.1:5000/ (Press CTRL+C to quit)
```

Figure 7.5 – The Flask server is started and ready to take requests

---

Note

We use Flask just for the demonstration purpose of creating client APIs. You might be interested in using the upgraded FastAPI python web framework to create the APIs.

---

3.   Now, we go to Postman and call the predict-online API, as shown in *Figure 7.6*:

Figure 7.6 – Calling the predict-online API from Postman

We send a request with the [[1, 1, 1], [2, 2, 2]] input data and get the response shown in the bottom panel of *Figure 7.6*:

4. If we check the console now, we will notice that the `print` statement inside the `update_model` function has appeared. From this execution, we get the following output:

```
Updating the model
New coefficients now
[0.29263309 0.29263309 0.29263309]
New intercept now
[0.16956489]
```

So, we notice the original coefficients have now been updated. The initial `[0.29263344 0.29263344 0.29263344]` coefficients are now changed to `[0.29263309 0.29263309 0.29263309]`, and the original `[0.16956488]` intercept is changed to `[0.16956489]`.

This example updates the model before sending the response to the user. However, we can send the response before updating the model. In that case, we can update the model asynchronously without blocking the response. We can also store a number of requests in a queue and update after a certain number of requests are stored. We can use a different service to update the model in that situation.

In this section, we have walked through a basic dummy end-to-end example of serving a model using the online serving pattern. You can take this workflow to serve any real-life model where the online serving pattern is a suitable solution.

We will conclude this chapter with a summary and point out some further reading.

Summary

In this chapter, we have discussed the online model serving pattern. We have introduced you to the definition of online model serving and different modes of online model serving. We have seen that, in online model serving, we try to keep the model updated with newly available data.

We then introduced some example cases where online model serving would be essential. In those cases, the model cannot tolerate delays in adapting to new data. The purpose of the model may be compromised if we do not update the model regularly with newly arrived data.

We then discussed some of the challenges in online model serving and introduced some ideas for how to address those challenges. We concluded the chapter with a dummy end-to-end example of online model serving.

In the next chapter, we will talk about two-phase model serving where two parallel versions of a model (one heavy, one light) are used to enable serving on low-memory and low-network devices.

Further reading

- To learn more about `partial_fit` in `scikit-learn`, you can follow this link: `https://scikit-learn.org/0.15/modules/scaling_strategies.html#incremental-learning`.

- To learn more about checkpointing DNN models to continue training with new data during an update, please follow this link: `https://machinelearningmastery.com/checkpoint-deep-learning-models-keras/`.

8

Two-Phase Model Serving

In this chapter, we will discuss the **two-phase prediction pattern**. In the two-phase prediction pattern, we deploy two different models. The bigger and more complex model is deployed on the server. In most cases, the users of this model are edge devices where the network may fluctuate. So, in the case of bad network access, an edge device can use a lightweight model to get predictions for basic use cases. For broader and more accurate predictions, the devices can get the prediction by calling APIs to the model deployed to the server. We will discuss the serving of models in this scenario of edge devices that exist in unstable networking conditions.

We will cover the following topics in this chapter:

- Introducing two-phase model serving
- Exploring two-phase model serving techniques
- Use cases of two-phase model serving

Technical requirements

In this chapter, we will use the TensorFlow library in addition to the libraries we have used before. We will be using the TensorFlow Lite converter to convert the large model to a smaller model. You should have Postman or another REST API client installed in order to send API calls and see the responses. All the code for this chapter is provided at `https://github.com/PacktPublishing/Machine-Learning-Model-Serving-Patterns-and-Best-Practices/tree/main/Chapter%208`.

If you get `ModuleNotFoundError` while trying to import any library, then you should install that module using `pip3 install <module_name>`. To find the exact versions of different modules, please look at the `requirements.txt` file in the repository.

Introducing two-phase model serving

In this section, we will discuss the basic concepts related to the two-phase model serving pattern.

The two-phase model serving pattern deploys two models for prediction. One large model is deployed on the distributed server, and one small model is deployed on the edge device. The large model is usually beyond the memory limit of the edge device and thus can't be deployed there. The smaller model deployed on the edge device is called the **phase one** model. The large model deployed on the cloud is known as the **phase two** model. This model is large and updated frequently to provide the most accurate predictions.

Two-phase model serving is very important when edge devices are involved in the overall system, and the predictions on these edge devices are essential, irrespective of the network conditions.

The phase one model is used for making predictions for two main reasons:

- To provide predictions if the device is offline. These predictions are not as accurate as the big model's predictions. However, these predictions are OK for getting an overview of the situation, such as getting the weather forecast or a route planner while offline. A diagram of two-phase serving in this scenario is given in *Figure 8.1*:

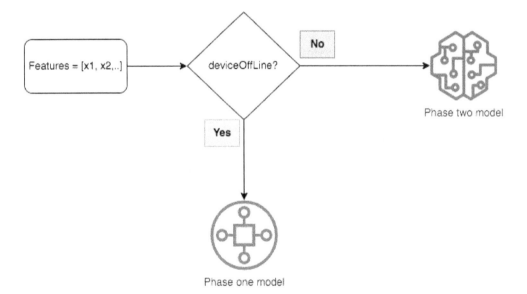

Figure 8.1 – Two-phase model serving scenario addressing offline state predictions

- To avoid calling APIs to the large model for unnecessary observations, we can keep calling the prediction API to the large model. However, this can be expensive. So, we call the phase one model to make a preliminary decision on whether we should call the API or not. For example, let's say we have a goal to predict the categories of tigers in a forest. We can have a phase one model in an edge device that will make a binary decision of whether the observed animal is a tiger or not. If the observation is yes, then we can send the features to the phase two model to determine the classification of the tiger's type. An example overview diagram of this serving is shown in *Figure 8.2*:

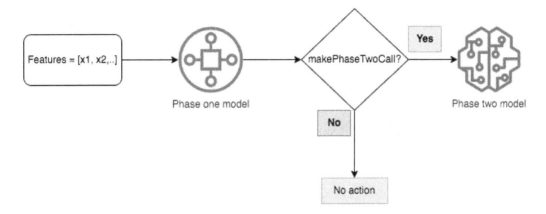

Figure 8.2 – High-level diagram of two-phase serving where the phase two model is called based on the prediction of the phase one model

In this section, we have seen an overview of two-phase model serving. In the next section, we will discuss the techniques required in two-phase model serving.

Exploring two-phase model serving techniques

Two-phase model serving can be one of the following three types, depending on the strength of the models:

- Quantized phase one model
- Separately trained phase one model with reduced features
- Separately trained different phase one and phase two models

Quantized phase one model

With this type, we first develop the phase two model to be deployed on the server. Then, we carry out integer quantization of the phase two model to form the phase one model. Integer quantization is an optimization technique that converts floating point numbers to 8-bit integer numbers. This way, the size of the model can decrease by a certain degree.

For example, if we convert 64-bit floating point numbers to 8-bit integers, we can get up to an 8-times reduction (64/8 = 8). A basic example of reducing the size of a floating point NumPy array to a uint8 NumPy array is shown in the following code block:

```
import numpy as np
import sys
X = [
    [1.5, 6.5, 7.5, 8.5],
    [1.5, 6.5, 7.5, 8.5],
    [1.5, 6.5, 7.5, 8.5],
    [1.5, 6.5, 7.5, 8.5]
]
A = np.array(X, dtype='float64')
B = A.astype(np.uint8)
print(A)
print(sys.getsizeof(A))
print(B)
print(sys.getsizeof(B))
```

In this code snippet, we take a NumPy array of floating point numbers and then convert the type of the array to uint8. We compute the size of the arrays in bytes before and after the conversion using sys.getsize. We get the following output from the code snippet:

```
[[1.5 6.5 7.5 8.5]
 [1.5 6.5 7.5 8.5]
 [1.5 6.5 7.5 8.5]
 [1.5 6.5 7.5 8.5]]
240
[[1 6 7 8]
 [1 6 7 8]
 [1 6 7 8]
 [1 6 7 8]]
128
```

So, we notice the floating-point numbers are all converted to integers, and the size of the array is reduced from 240 bytes to 128 bytes after the conversion.

The weights and biases in a deep neural network are represented by 32-bit floating point numbers. Therefore, after 8-bit integer quantization, we can reduce the size of the model by 32/8 = 4 times.

To demonstrate the reduction in the size of the model after quantization, we will use the example at https://www.tensorflow.org/lite/performance/post_training_integer_quant.

We will train a model for the MNIST dataset and then save it. We will compute the size of the saved model. Then we will do full integer quantization and save the quantized model, then finally, we will compute the size of the quantized model.

Training and saving an MNIST model

In this section, we will train a basic deep-learning model for recognizing the MNIST digits. We will follow the following steps to train and save the trained model:

1. First of all, we download the MNIST data using the following code snippet:

    ```
    mnist = tf.keras.datasets.mnist
    (train_images, train_labels), (test_images, test_labels)
    = mnist.load_data()
    ```

2. Then, we normalize the data using the following code snippet:

    ```
    train_images = train_images.astype(np.float32) / 255.0
    test_images = test_images.astype(np.float32) / 255.0
    ```

3. Then we define the sequential model structure as follows:

    ```
    model = tf.keras.Sequential([
      tf.keras.layers.InputLayer(input_shape=(28, 28)),
      tf.keras.layers.Reshape(target_shape=(28, 28, 1)),
      tf.keras.layers.Conv2D(filters=12, kernel_size=(3, 3),
    activation='relu'),
      tf.keras.layers.MaxPooling2D(pool_size=(2, 2)),
      tf.keras.layers.Flatten(),
      tf.keras.layers.Dense(10)
    ])
    ```

    We can see that the model has six layers.

4.   Then we compile and train the model using the following code snippet:

```
model.compile(optimizer='adam',
               loss=tf.keras.losses.
SparseCategoricalCrossentropy(
                   from_logits=True),
               metrics=['accuracy'])
model.fit(
   train_images,
   train_labels,
   epochs=5,
   validation_data=(test_images, test_labels)
)
```

Notice the final accuracy of the model in the terminal as the model finishes training:

```
Epoch 5/5
1875/1875 [==============================] - 15s 8ms/step
- loss: 0.0602 - accuracy: 0.9826 - val_loss: 0.0633 -
val_accuracy: 0.9798
```

We see that the model has ~98% accuracy in detecting the digits from the MNIST dataset.

5.   We save the model using `model.save("mnist")`. We will see that the model is saved to the `mnist` directory, and the directory structure is shown in *Figure 8.3*:

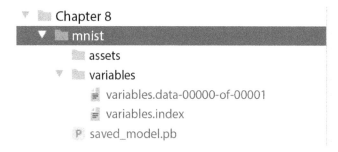

Figure 8.3 – Directory structure of the saved neural network to detect MNIST digits

Full integer quantization of the model and saving the converted model

Now we will do the full integer quantization of the model. We will use Tensorflow Lite converter `tf.lite.TFLiteConverter.from_keras_model` to perform the full integer quantization of the model.

1. First, we create a converter using the following code snippet:

    ```
    converter = tf.lite.TFLiteConverter.from_keras_
    model(model)
    ```

    This code creates a converter using `tf.lite.TFLiteConverter.from_keras_model` and it takes the model that we created earlier as a parameter.

2. We are going to do full integer quantization, so we need to specify the representative dataset for the converter. We create the representative dataset from the training data in the following way and assign that representative data to the `representative_dataset` property of the converter:

    ```
    def representative_data_gen():
      for input_value in tf.data.Dataset.from_tensor_
    slices(train_images).batch(1).take(100):
        yield [input_value]
    converter.representative_dataset = representative_data_
    gen
    ```

3. Now we set the optimization strategy and the types of conversion as follows:

    ```
    converter.optimizations = [tf.lite.Optimize.DEFAULT]
    converter.target_spec.supported_ops = [tf.lite.OpsSet.
    TFLITE_BUILTINS_INT8]
    converter.inference_input_type = tf.uint8
    converter.inference_output_type = tf.uint8
    ```

    We have set the optimizations to the default `tf.lite` optimization strategy. We have also set the `inference_input_type` and `inference_output_type` values to `tf.uint8`.

4. We can now convert the model to the quantized model using the following command:

    ```
    tflite_model_quant = converter.convert()
    ```

    We have now converted the model and stored the quantized and reduced model to the `tflite_model_quant` variable.

5. We can save the quantized model to our local directory using the following code snippet:

    ```
    tflite_models_dir = pathlib.Path("quantized_model/")
    tflite_model_quant_file = tflite_models_dir/"mnist_model_
    ```

```
quant.tflite"
tflite_model_quant_file.write_bytes(tflite_model_quant)
```

We have saved the model in the `quantized_model` directory. If the directory does not exist, you might need to create the directory in your project folder before running this code.

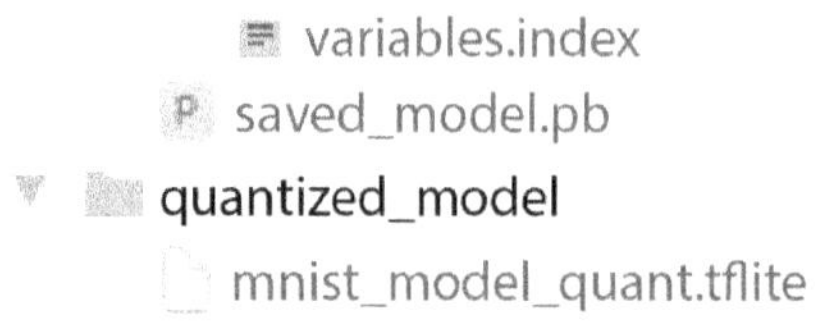

≣ variables.index

ℙ saved_model.pb

▾ ▨ quantized_model

mnist_model_quant.tflite

Figure 8.4 – Directory structure of the saved quantized model

The model is saved with the name `mnist_model_quant.tflite`. The directory structure of the saved model is shown in *Figure 8.4*.

Comparing the size and accuracy of the models

We now have two models. One model is the large model where the weights and biases are in the `float32` form. This is used as the phase two model, which will be deployed on the server. In our example, we saved the model with the name `saved_model.pb`. The other model is the quantized model, where the weights and biases use the `uint8` types. Let's compare the sizes of the two models:

1. Let's create a separate program with the following code snippet:

    ```
    import os
    size = os.stat('mnist/saved_model.pb').st_size
    print("Size of phase two model ", size)
    size2 = os.stat('quantized_model/mnist_model_quant.
    tflite').st_size
    print("Size of phase one model ", size2)
    reduction = round(size/size2, 2)
    print("Reduction of size is ", reduction, "times")
    ```

    From this code snippet, we get the following output:

    ```
    Size of phase two model   97916
    Size of phase one model   24576
    Reduction of size is   3.98 times
    ```

    The quantized model is almost four times smaller. However, the phase two model has some other artifacts, such as variables and assets, along with the model, as shown in *Figure 8.3*. We can check the sizes of these variables by going inside the directory. The size of these variables is shown in *Figure 8.5*. We can see that the variables are nearly three times the size of the model. Therefore, the actual reduction after quantization is much more than we got by comparing the model file sizes.

▼ 📁 assets	Aug 11, 2022, 8:38 AM	--	Folder
📄 saved_model.pb	Aug 12, 2022, 11:01 PM	98 KB	Document
▼ 📁 variables	Aug 12, 2022, 11:01 PM	--	Folder
📄 variables.data-00000-of-00001	Aug 12, 2022, 11:01 PM	249 KB	Document
📄 variables.index	Aug 12, 2022, 11:01 PM	1 KB	Document

Figure 8.5 – The sizes of the items in the variables folder for the phase two model

The MNIST model is simple, so the sizes are small. For complex models, the sizes can be very big.

2. To evaluate and get predictions from the model, we first need to create an interpreter and get predictions from the interpreter. We compute the predictions from the quantized model with the following code for each of the test images:

```
test_image_indices = range(test_images.shape[0])
interpreter = tf.lite.Interpreter(model_path=str(tflite_
model_quant_file))
interpreter.allocate_tensors()
input_details = interpreter.get_input_details()[0]
output_details = interpreter.get_output_details()[0]
predictions = np.zeros((len(test_image_indices),),
dtype=int)
for i, test_image_index in enumerate(test_image_indices):
    test_image = test_images[test_image_index]
    test_label = test_labels[test_image_index]

    # Check if the input type is quantized, then rescale
input data to uint8
    if input_details['dtype'] == np.uint8:
      input_scale, input_zero_point = input_
details["quantization"]
      test_image = test_image / input_scale + input_zero_
point

    test_image = np.expand_dims(test_image, axis=0).
astype(input_details["dtype"])
```

```
    interpreter.set_tensor(input_details["index"], test_
image)
    interpreter.invoke()
    output = interpreter.get_tensor(output_
details["index"])[0]
    predictions[i] = output.argmax()
```

In this code snippet, we create the interpreter from the quantized model with `interpreter = tf.lite.Interpreter(model_path=str(tflite_model_quant_file))`. We then rescale the input images to pass through the full-integer quantized model using `test_image = test_image / input_scale + input_zero_point`. Then we set the rescaled image as input to the model with `interpreter.set_tensor(input_details["index"], test_image)`. Then we make a prediction using `interpreter.invoke()`. We then get the prediction result from `output = interpreter.get_tensor(output_details["index"])[0]` and take the maximum value of the 10 output tensors as the prediction. We repeat this process for all the test images to get the predictions for all of them.

3.   Then we use the predictions to compute the accuracy in the following code snippet:

```
accuracy = (np.sum(test_labels== predictions) * 100) /
len(test_images)

print('Quantized model accuracy is %.4f%% (Number of test
samples=%d)' % (accuracy, len(test_images)))
```

The output from the `print` statement is as follows:

```
Quantized model accuracy is 97.9300% (Number of test
samples=10000)
```

In this quantized model, the accuracy is close to the accuracy we got from the bigger phase two model. However, depending on the reduction goal and the model's complexity, we might get much lower accuracy.

In this subsection, we have discussed the type of two-phase model serving where the phase one model is quantized. To find out more about different quantization strategies, please visit `https://www.tensorflow.org/model_optimization/guide/quantization/training`.

In the next subsection, we will discuss the two-phase model serving, where the phase one model is trained separately with fewer features.

Separately trained phase one model with reduced features

Sometimes we can train two different models for phase one and phase two deployments. For the training of the phase one model, we can use only a few of the most important features selected using the built-in feature selection strategies provided by the **machine learning** (**ML**) libraries such as Tensorflow, and scikit-learn. In this way, we can reduce the size of the model by using fewer features.

We will show a very basic example where we select the top feature from the `iris` dataset (https://archive.ics.uci.edu/ml/datasets/iris) and train a phase one logistic regression model with that one feature to classify the type of iris plants based on their physical features. We will train the phase two model with all the features:

1. First of all, we train the phase two model using the following code snippet:

    ```
    X, y = load_iris(return_X_y=True)
    print("Original data dimensions:", X.shape)
    phase_one_model = LogisticRegression(random_state=0, max_
    iter=1000).fit(X, y)
    score = phase_one_model.score(X, y)
    print("Accuracy of the phase two model:", score)
    p = pickle.dumps(phase_one_model)
    print("Size of the phase two model in bytes:", sys.
    getsizeof(p))
    ```

    In this code, we fit the `LogisticRegression` model with the full dataset. We then compute the accuracy score and the size of the model. We get the following output from this code snippet:

    ```
    Original data dimensions: (150, 4)
    Accuracy of the phase two model: 0.9733333333333334
    Size of the phase two model in bytes: 851
    ```

    The input data has 150 rows and 4 features. All the features are used to train the phase two model. We got an accuracy of ~97.33%, and the size of the model is 851 bytes.

2. Now, we will separately train the phase one model. We will only take a single feature out of the four features from the dataset. We will select the best feature using the `SelectKBest` method from scikit-learn. The following code snippet computes the most important feature and trains the model with that most important feature:

    ```
    X_new = SelectKBest(chi2, k=1).fit_transform(X, y)
    print("Data dimensions after feature reduction:", X_new.
    shape)
    ```

```
phase_two_model = LogisticRegression(random_state=0, max_
iter=1000).fit(X_new, y)
score = phase_two_model.score(X_new, y)
print("Accuracy of the phase one model: ", score)
p = pickle.dumps(phase_two_model)
print("Size of the phase one model in bytes:", sys.
getsizeof(p))
```

We have only selected one feature using k=1 in the SelectKBest method. Then we train another LogisticRegression model using the X_new dataset with a single feature.

From this code snippet, we get the following output:

```
Data dimensions after feature reduction: (150, 1)
Accuracy of the phase one model:  0.9533333333333334
Size of the phase one model in bytes: 779
```

The accuracy of the phase one model is ~95.33%, which is ~2% less than the phase two model. And the size of the model is 779 bytes, which is 851 – 779 = 72 bytes less than the phase two model. The data dimension is now (150, 1), which means we used all 150 rows but only one feature to train this new model.

The problem with training the phase one model in this way is that the model is underfitted, so we get more errors.

We can deploy this reduced model to an edge device and the phase two model to a server to follow the two-phase model serving pattern.

Separately trained different models

In this case, we train different models for phase one and phase two. Usually, the phase one model, which is deployed on the edge device, makes a binary prediction to decide whether the phase two model needs to be called or not. Examples of different scenarios include the following:

- The phase one model predicts whether there is congestion on a road. If there is congestion, the data is sent to the phase two model to update the estimated arrival time from a distance monitoring app.

- The phase one model can determine whether a particular disease is present in a patient. If the disease is present, then the data can be sent to the phase two model to get a detailed analysis of the severity of the disease.

Besides these examples, there are many other scenarios where the phase one model can be very simple to make a binary decision to invoke the phase two complex model.

To demonstrate the different phase one and phase two models, let's consider a dummy scenario. The phase one model will return `True` if the input is less than 0.5; otherwise, it will return `False`. If the phase one model returns `True`, only then will the phase two model be called. The phase two dummy model simply makes a random choice and returns that as its prediction. The example is shown in the following code snippet:

```python
def predict_phase_one_model(x):
    print("The value of x is", x)
    if x < 0.5:
        return True
    else:
        return False

def predict_phase_two_model():
    print("Phase two model is called")
    prediction = np.random.choice(["Class A", "Class B", "Class C"])
    return prediction

if __name__=="__main__":
    phase_one_prediction = predict_phase_one_model(random.uniform(0, 1))
    if phase_one_prediction == True:
        response = predict_phase_two_model()
        print(response)
    else:
        print("Phase two model is not called")
```

We only call the phase two model inside the `if` block, given the prediction from the phase one model is `True`. The output from two separate runs of the program is as follows:

- Run 1:

  ```
  The value of x is 0.05055747049395931
  Phase two model is called
  Class A
  ```

- Run 2:

```
The value of x is 0.7624742118541643
Phase two model is not called
```

We can see that the phase two model is called in Run 1 as the value of the input is less than 0.5, causing the phase one model to return `True`. The phase two model is not called in Run 2 because the phase one model returns `False`.

We have used a dummy example to demonstrate the workflow of calling the phase two model based on the prediction from the phase one model. This scenario can help us to avoid making unnecessary API calls from the edge device to the phase one model deployed on the server.

For deep learning, we can also separately train two different models. The phase one model will be simple, and the phase two model will be complex. For example, for the MNIST dataset phase two model, we will train a VGG-16 model with 16 layers. We will use the model from `https://www.kaggle.com/code/chandraroy/vgg-16-mnist-classification` to create the VGG-16 model.

After creating the model, we print the model summary:

```
Layer (type)                Output Shape              Param #
=================================================================
conv2d (Conv2D)             (None, 28, 28, 32)        320


conv2d_1 (Conv2D)           (None, 28, 28, 64)        18496


... Truncated output ...
dense_2 (Dense)             (None, 10)                40970
=================================================================
Total params: 33,624,202
Trainable params: 33,621,258
Non-trainable params: 2,944
```

We have truncated the output to make it easier to read. To view the complete output, please run the `mnist_vgg.py` file from the Chapter 8 folder in the book's code repository. The link to the code is provided in the *Technical requirements* section. As you can see, the model has more than 33 million parameters.

We can create a simple sequential model for phase one using the following code snippet:

```
model = tf.keras.Sequential([
  tf.keras.layers.InputLayer(input_shape=(28, 28)),
  tf.keras.layers.Reshape(target_shape=(28, 28, 1)),
```

```
    tf.keras.layers.Conv2D(filters=12, kernel_size=(3, 3),
activation='relu'),
    tf.keras.layers.MaxPooling2D(pool_size=(2, 2)),
    tf.keras.layers.Flatten(),
    tf.keras.layers.Dense(10)
])
model.summary()
```

The output from `model.summary()` is as follows:

```
Layer (type)                    Output Shape              Param #
=================================================================
reshape (Reshape)               (None, 28, 28, 1)         0

conv2d (Conv2D)                 (None, 26, 26, 12)        120

max_pooling2d (MaxPooling2D)    (None, 13, 13, 12)        0

flatten (Flatten)               (None, 2028)              0

dense (Dense)                   (None, 10)                20290
=================================================================
Total params: 20,410
Trainable params: 20,410
Non-trainable params: 0
```

This model has only 20,410 parameters. We can use this small model as the phase one model and the complex and large VGG-16 model as the phase two model.

The phase two model size is shown in *Figure 8.6*. The size of the variables is more than 400 MB, whereas the size of the variables in the phase one basic model is ~200 KB, as shown in *Figure 8.5*:

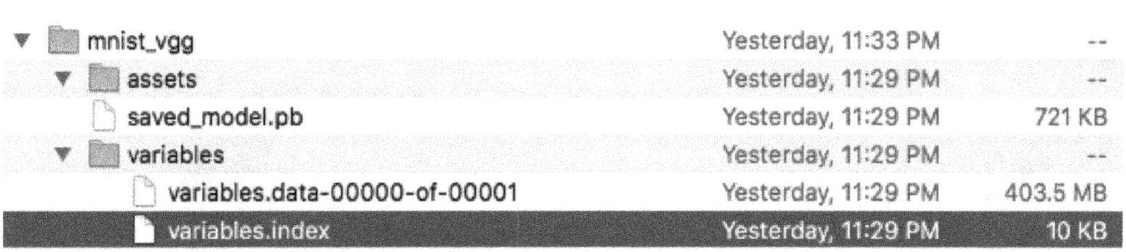

▼ 📁 mnist_vgg	Yesterday, 11:33 PM	--
▼ 📁 assets	Yesterday, 11:29 PM	--
📄 saved_model.pb	Yesterday, 11:29 PM	721 KB
▼ 📁 variables	Yesterday, 11:29 PM	--
📄 variables.data-00000-of-00001	Yesterday, 11:29 PM	403.5 MB
📄 variables.index	Yesterday, 11:29 PM	10 KB

Figure 8.6 – Size of the trained VGG-16 model for the MNIST dataset

In this section, we have discussed some technical approaches and ideas for the two-phase prediction pattern. We have discussed different combinations of phase one and phase two models to set up the two-phase prediction pattern. In the next section, we will discuss some use cases of the two-phase model serving pattern.

Use cases of two-phase model serving

In this section, we will discuss some example cases where two-phase model serving can be used.

Case 1 – a fitness tracking device

Imagine that there is an app on our handheld device that recommends fitness tasks based on our activity. The device will go offline due to the absence of the network in a remote area or due to a poor signal. In that case, the device can't make the API calls to the remote server to get recommendations. Therefore, we need a model deployed on the device that will be smaller than the model on the server. This small model will serve as the phase one model. A fitness tracker device can collect different features and use those features to get detailed recommendations from the server. However, the phase one model can only use a few critical features denoting whether the blood oxygen level is low, whether the heart rate is abnormal, whether the user has had enough rest, and so on. Based on these critical features, the device can then provide critical recommendations on whether the user needs to rest, drink water, and so on. The device can check the network strength whenever a recommendation needs to be made. If the network is weak, the device will ask the phase one model to make recommendations, and if the network is strong, the device will make an API call to get recommendations from the phase two model on the server.

Case 2 – a location with low internet speed

The internet is slow in many countries. In this scenario, we can't keep calling the APIs on the server all the time to get predictions. We need to use the offline model on the device to avoid making API calls. In that case, we have to update the offline phase one model frequently. We can use both the phase two model and the phase one model to make predictions and then compute the error. We can use this error to improve the phase one model and also to provide an error estimation to the user with respect to the phase two model. For example, here are some dummy predictions from the phase one and phase two models:

X1	X2	Phase one prediction (P1)	Phase two prediction (P2)	% Deviation 100*(P2 – P1)/ P2
2	3	6	6.5	7.69%
5	6	7	7.5	6.67%
7	8	8	8.5	5.88%
9	10	9	9.5	5.26%
11	12	10	10.5	4.76%

Figure 8.7 – Predictions from the phase one and phase two models for some dummy samples

Now we can compute the median or average deviation from all the deviations in the last column and then use that metric, along with the predictions from the phase one model, to provide error estimates to the users.

For example, we can compute the median and average deviations using the following code snippet:

```
import numpy as np

deviations = np.array([7.69, 6.67, 5.88, 5.26, 4.76])
med_deviations = np.median(deviations)
print(med_deviations)

avg_deviations = np.average(deviations)
print(avg_deviations)
```

We get the following output:

```
5.88
6.052
```

Whenever we provide predictions from the phase one model, the predictions can be tagged with these error estimates. For example, if the model predicts 6, we can state that the prediction is likely to have a deviation of 6.052%, the average deviation. So, the most likely actual value is as follows:

$6 / (100 – 6.052)\% = 6 / 0.93948 \sim 6.39$

Interestingly, 6.39 is closer to the value of **6.5** predicted by the phase two model.

The high-level strategy of combining the predictions from the phase one model and the error estimation model is shown in *Figure 8.8*:

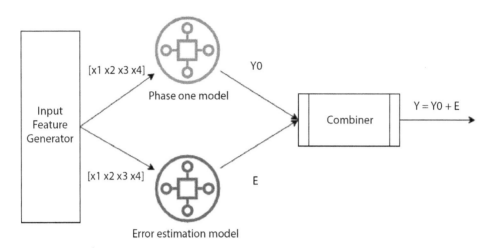

Figure 8.8 – Combining the predictions from the phase one model and the error estimator model to get a more accurate prediction

In *Figure 8.8*, we have two models. The **Phase one model** and the **Error estimation model** both are passed the same input. The Error estimation model is trained using the deviation in prediction data in *Figure 8.7*. The predictions from the models are combined inside the **Combiner** block, and we get the final prediction **Y**, which is expected to be closer to the phase two prediction model.

Case 3 – weather forecasts

For weather forecasts from apps on edge devices, we might have to depend on the offline phase one model to get an estimated prediction of the weather. In this case, the phase one forecasting model will use features tracked using sensors on the device. For example, we can collect features such as air temperature, wind speed, luminosity, and humidity. The phase one model can be created to use only these critical features. On the other hand, the phase two model can use features from weather stations as well as features from edge devices.

Case 4 – route planners

We use edge devices to make route planners for our journeys. We want these apps to provide recommendations even if the network is offline. A phase one model may be used to provide the best possible offline route. This phase one model might not be able to consider live information on congestion, crashes, and so on. However, it can use some primary basic features such as distance, geolocation, source location, and destination location to provide a route to the destination, so that you do not get lost in remote locations where the network is very weak.

Case 5 – smart homes

In smart homes, we can have devices that set the room temperature, start sprinklers, turn on air conditioning, and so on. In these cases, we can have two different models. The first model will listen to the basic queries that will enable the default service or turn off the service.

For example, for commands such as the following, we can use a phase one model:

- Turn on the light
- Turn off the light
- Turn on the fan
- Turn off the fan
- Turn on the air conditioning
- Turn off the air conditioning

These commands can be addressed even by a simple model that can make binary *YES/NO* binary decisions.

Now let's consider the following detailed commands that need more complex analysis:

- Turn on the light in the master bedroom
- Turn on the light in bathroom one
- Turn on the AC and set the temperature to T
- Turn on the fans in the living room

These commands can't be handled easily by the small phase one model. So, we can redirect the prediction of these commands to the phase two model deployed in the cloud.

In this section, we have discussed some use cases of two-phase model serving. Besides the use cases described here, there are many other use cases where two-phase model serving should be chosen as the best option.

Summary

In this chapter, we have discussed two-phase model serving. We have explained what two-phase model serving is and why it is needed. We have also discussed different combinations of phase one and phase two models. We have seen that the phase one model can be created via quantization of the phase two model, which involves training only a single model. The phase one model can also be trained separately from the phase two model. These techniques are discussed along with some basic examples throughout the chapter. We have also discussed some examples of two-phase model serving.

In the next chapter, we will talk about the pipeline pattern. We will learn how ML pipelines are created, how different stages in the pipeline are interconnected to serve the model, and how the execution of the pipeline is scheduled.

Further reading

- To find out more about TensorFlow Lite, visit `https://www.tensorflow.org/lite`

- To find out more about model optimization and quantization techniques, visit `https://www.tensorflow.org/lite/performance/model_optimization`

- To find out more about the VGG-16 model for MNIST, visit `https://www.kaggle.com/code/chandraroy/vgg-16-mnist-classification`

9

Pipeline Pattern Model Serving

In this chapter, we will discuss the **pipeline pattern** of model serving. In this pattern, we create a pipeline with a number of stages. In each of the stages, some functionalities (such as data collection and feature extraction) are performed. The pipeline is started using a scheduler or it can be started manually. We will discuss how to create a pipeline for serving **machine learning** (**ML**) models and demonstrate how the pipeline pattern can be used for this purpose. The high-level topics that will be covered in this chapter are as follows:

- Introducing the pipeline pattern

- Introducing Apache Airflow

- Demonstrating a machine learning pipeline using Apache Airflow

- Advantages and disadvantages of the pipeline pattern

Technical requirements

In this chapter, we will use Apache Airflow to demonstrate pipeline creation via the pipeline pattern. For all the resources, documentation, and tutorials on Apache Airflow, please visit their website: `https://airflow.apache.org/`. We will give a high-level overview of Apache Airflow and how to use it to create pipelines, but we will not cover details of the Apache Airflow infrastructure. Feel free to learn more about Apache Airflow to see how to create complex pipelines and workflows. Please install Apache Airflow following the instructions for your OS on their official website.

All the code for this chapter is provided at `https://github.com/PacktPublishing/Machine-Learning-Model-Serving-Patterns-and-Best-Practices/tree/main/Chapter%209`.

If you get `ModuleNotFoundError` while trying to import any library, then you should install that module using the `pip3 install <module_name>` command.

Introducing the pipeline pattern

In the pipeline pattern, we create a pipeline to serve an ML model. Instead of all the steps of an ML pipeline happening in a central place, the steps are separated to occur in different stages of a pipeline. A pipeline is a collection of a number of stages in the form of a **directed acyclic graph** (**DAG**). We will start by describing DAGs in the following subsection and then gradually describe how the pipeline is formed by connecting different stages as a DAG.

A DAG

A DAG is a graph where the vertices are connected by directed edges, and no path in the graph forms a cycle. A path is a sequence of edges in a graph. A path forms a cycle if the start and end nodes of the path are the same. The directions of the edges are denoted by arrows. Nodes in the DAG are represented by circles. To understand DAGs graphically, let's see a few graphs that are not DAGs.

First of all, let's look at the graph in *Figure 9.1*. This graph is not a DAG as one of the edges of this graph, **AC**, does not have a direction.

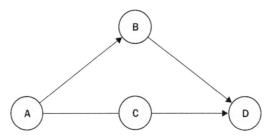

Figure 9.1 – A non-DAG as the edge AC does not have a direction

Therefore, the presence of direction in a graph is a key requirement for the graph to be a DAG. This is evident from the name of this kind of graph. The letter *D* denotes *Directed*, which can help you to remember that the edges in the graph will have direction.

Now let's look at the graph in *Figure 9.2*. In this graph, again, all the edges have direction. However, the graph is still not a DAG.

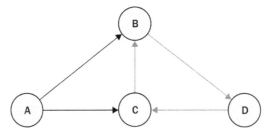

Figure 9.2 – A non-DAG due to the presence of a cyclic path, CBD

The second letter *A* in DAG denotes *Acyclic*, which means there can't be any cyclic paths in the graph. The graph in *Figure 9.2* has a cyclic path, **CBD**, which is colored red in the figure. Due to the presence of this cyclic path, the graph in *Figure 9.2* is *not* a DAG. In this subsection, we have discussed some concepts on DAG. In the next subsection, we will discuss the stages of the pipeline and how a stage can be represented as a node in the DAG.

Stages of the machine learning pipeline

A stage of a pipeline is like a node in a graph. In these stages, a unit of activity happens. We can split the bigger task of the end-to-end process of serving the ML model, starting with data collection, into multiple steps as follows:

1. **Data collection**: In this stage, we collect data for the ML model from different data sources. We can collect data from surveys, sensors, social media posts, and so on.

2. **Data cleaning**: After collecting the data, we need to do some cleaning of the data to make it usable in the program. The data cleaning process can involve some manual operations as well as using some built-in functions and tools.

3. **Feature extraction**: In this step, we extract the features from the cleaned data to be used for training the model.

4. **Training the model**: After the feature extraction, we can train the model using those features. If we need more data or more features we can go back to the previous stages.

5. **Testing the model**: After the model is trained, we test the performance of the model with some unseen data.

6. **Saving the model**: After the testing is done and we are happy with the performance of the model, we can save the model for deploying or serving.

> **Note**
> Please note that here serving is not decoupled from the pipeline. However, the main goal of serving is to provide customers with access to the model's response, usually from the last stage of the pipeline. We can take the prediction created from the prediction stage directly or create some REST APIs to access the predictions.

The pipeline is the connection of stages in the form of a DAG. The graph needs to be acyclic because we do not want to get lost in a loop inside the pipeline. If a cyclic path exists in the pipeline, then there is a chance that the pipeline will keep looping through that cyclic path. Therefore, we will use a DAG to create the pipeline. Using a DAG will ensure the following things:

- The directions of the edges will provide clear guidelines on how to move forward from a stage

- Acyclic paths will ensure that our program does not get trapped in a never-ending loop

In this section, we have learned about the pipeline pattern, DAGs, and how DAGs can be used to create pipelines via the pipeline pattern. In the next section, we will introduce you to Apache Airflow.

Introducing Apache Airflow

In this section, we will give you a high-level overview of **Apache Airflow**. To find out more details about Apache Airflow, read the documentation on their official website. The link to the official website is provided in the *Technical requirements* section.

Getting started with Apache Airflow

Apache Airflow is a full stack platform for creating workflows or pipelines using Python, scheduling the pipeline, and also monitoring the pipeline using the GUI dashboard provided by the platform. To see the tool on your local machine and understand how to create a pipeline, please install Apache Airflow. I am showing the steps I used to install Apache Airflow on macOS:

1. To install Apache Airflow, first, create a directory on your local machine and set the AIRFLOW_ HOME variable. This is important because Apache Airflow will install the configurations here and fetch the workflows from this directory. I created a directory on my desktop and exported the AIRFLOW_HOME path using the following command from the terminal:

    ```
    export AIRFLOW_HOME=/Users/mislam/Desktop/airflow-
    examples
    ```

2. Then, you can install Apache Airflow using the following command:

    ```
    pip install "apache-airflow[celery]==2.3.4" --constraint
    "https://raw.githubusercontent.com/apache/airflow/
    constraints-2.3.4/constraints-3.7.txt"
    ```

    You can also install Apace Airflow in a virtual environment instead of on your own platform. Follow the instructions on the installation page, https://airflow.apache.org/ docs/apache-airflow/stable/installation/index.html, to get detailed instructions on installation to your appropriate device.

3. Then we confirm whether the installation was successful by typing the following command:

    ```
    $ airflow version
    ```

    You should see the version number of Apache Airflow on the terminal. You can then go to the directory, which is set as AIRFLOW_HOME, and see some files created for you, as shown in *Figure 9.3*. You will not see a dags folder at the beginning. This is because you need to create it manually.

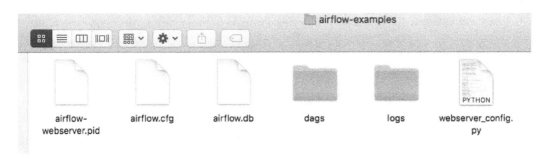

Figure 9.3 – Apache Airflow created installation files

We have to create the dags folder manually and place all the DAGs in it. To understand why we need to place all the DAGs in this folder, open the airflow.cfg file and find out the value for the dags_folder variable, as shown in *Figure 9.4*. Airflow fetches the DAGs from this folder. If you need to place DAGs from a different folder, you can change this path.

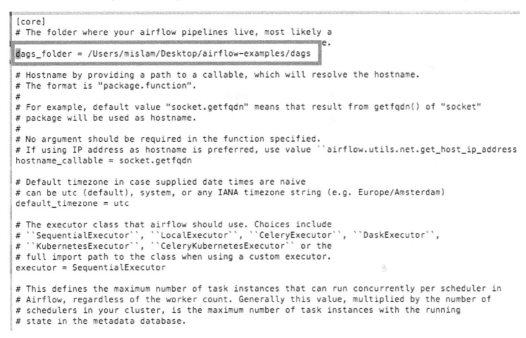

Figure 9.4 – The value of the dags_folder variable in the airflow.cfg file

Therefore, whenever we want to run a workflow and monitor it using the GUI, we need to place it inside the $AIRFLOW_HOME/dags folder.

Then, in the terminal, we run the following command to initialize the Airflow database that will be used to keep track of the users and the workflows:

```
airflow db init
```

4. Then, we need to create an Admin user to log in to the Airflow UI. We can create a dummy user using the following command:

```
airflow users  create --role Admin --username admin
--email admin --firstname admin --lastname admin
--password admin
```

Feel free to change the information if you want. We use the dummy values to get access to the UI for demonstration purposes.

Creating and starting a pipeline using Apache Airflow

After Apache Airflow is installed successfully, we can create workflows and see whether it is working:

1. Now, let's create a basic workflow using the following code snippet:

```
# The DAG object; we'll need this to instantiate a DAG
from datetime import datetime, timedelta
from airflow import DAG
from airflow.operators.bash import BashOperator

with DAG(
    'example_tutorial',
    default_args={
        'depends_on_past': False,
        'email': ['airflow@example.com'],
        'email_on_failure': False,
        'email_on_retry': False,
        'retries': 1,
        'retry_delay': timedelta(minutes=5)
    },
    description='A simple tutorial DAG',
    schedule_interval=timedelta(days=1),
    start_date=datetime(2022, 1, 1),
```

```
    catchup=False,
    tags=['dummy'],
) as dag:

    # t1, t2 examples of tasks created by instantiating
operators
    t1 = BashOperator(
        task_id='stage1',
        bash_command='python3 /Users/mislam/Desktop/
airflow-examples/dags/stage1.py',
    )


    t2 = BashOperator(
        task_id='stage2',
        bash_command=' python3 /Users/mislam/Desktop/
airflow-examples/dags/stage2.py',
    )
    t1 >> t2
```

In this code snippet, we first create a DAG using the DAG class provided by Airflow. We set the id of the DAG to example_tutorial. We also pass other properties, such as tags = ['dummy'], schedule_interval = timedelta(days=1), and so on. Using tags, we can search the workflow from the dashboard, as shown in *Figure 9.7*.

Then, we use BashOperator (provided by Airflow) to create a stage in the pipeline. According to Airflow terminology, stages are known as tasks. We create two tasks, t1 and t2, with IDs of stage1 and stage2, respectively. In the first stage, we run a Python file, stage1.py, and in the second stage, we run another Python file, stage2.py.

The stage1.py file contains the following code snippet:

```
import time
print("I am at stage 1")
time.sleep(5)
print("After 5 seconds I am existing stage 1")
```

And the stage2.py file contains the following code snippet:

```
print("I am at stage 2")
```

2. Now, we place stage1.py and stage2.py along with the file containing the DAG in the $AIRFLOW_HOME/dags directory.

3. Then we have to start `airflow scheduler`. This scheduler takes care of loading the new workflows to the UI, starting workflows when scheduled, and so on. We can start the scheduler using the following command:

    `airflow scheduler`

4. Then, let's start the Airflow web server using the following command:

    `airflow webserver`

    We should see that the web server has been started, as shown in *Figure 9.5*:

```
Johiruls-Air:Desktop mislam$ airflow webserver

     ____    |__( )_____  __/__  /_____      __
 ____ /| |_  /__  ___/_  /_ __  /_  __ \_  __ \_  __ \ | /| / /
 ___ ___ |  / _  /   _  __/  / / / / /_/ /_  / / /_  / // /|/ /
 _/_/  |_/_/  /_/    /_/    /_/  \____/____/ /_/ \__/|__/
Running the Gunicorn Server with:
Workers: 4 sync
Host: 0.0.0.0:8080
Timeout: 120
Logfiles: - -
Access Logformat:
=================================================================
[2022-09-02 18:14:37 -0500] [58676] [INFO] Listening at: http://0.0.0.0:8080 (58676)
[2022-09-02 18:14:37 -0500] [58778] [INFO] Booting worker with pid: 58778
[2022-09-02 18:14:37 -0500] [58779] [INFO] Booting worker with pid: 58779
[2022-09-02 18:14:37 -0500] [58780] [INFO] Booting worker with pid: 58780
[2022-09-02 18:14:37 -0500] [58781] [INFO] Booting worker with pid: 58781
```

Figure 9.5 – The Airflow web server has been started and is listening
at http://0.0.0.0:8080 or http://localhost:8080

5. Let's open a browser and visit `http://localhost:8080/`, and we will see the screen shown in *Figure 9.6*:

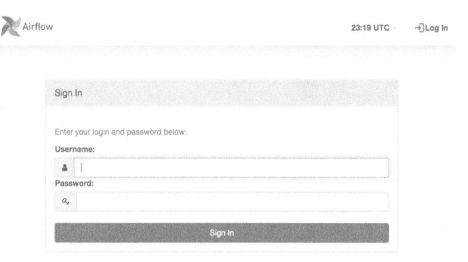

Figure 9.6 – The Sign In screen of the Airflow UI launched after visiting the URL http://localhost:8080/

6. After the **Sign In** screen appears, we have to enter the admin username and password that we created in the last sub-section. After successfully signing in, we will see a home screen with a list of workflows, as shown in *Figure 9.7*:

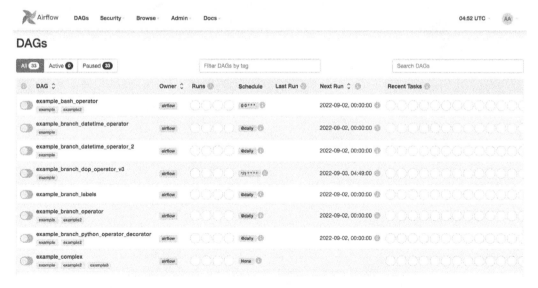

Figure 9.7 – List of workflows shown in the Airflow UI after signing in

7. We see the option to search the workflows in the two text boxes above the list. Let's filter the workflow by our dummy tag, and click on the DAG ID. It will take us to the page showing the details about the workflow. The workflow or pipeline is shown in *Figure 9.8*:

Figure 9.8 – Detailed view of the workflow first_airflow_pipeline

The home page of the workflow has a lot of information. You can view the history of runs, the next run time, the logs, the graphical representation of the pipeline, and more. For example, in our example, we created a pipeline with two stages, **stage1** and **stage2**. We see the pipeline on the dashboard, along with the direction of connection between the edges. This direction is specified in the code. In our code, we created the pipeline and specified the direction from **stage1** to **stage2** via t1 >> t2 in the following code snippet:

```
t1 = BashOperator(
        task_id='stage1',
        bash_command=' python3 /Users/mislam/Desktop/
airflow-examples/dags/stage1.py',
    )

    t2 = BashOperator(
        task_id='stage2',
        bash_command=' python3 /Users/mislam/Desktop/
airflow-examples/dags/stage2.py',
    )
    t1 >> t2
```

If we want to reverse the direction, then we have to replace t1 >> t2 with t2 >> t1. The >> or << operator, which looks like an arrowhead, gets converted to an arrow in the pipeline, and the operator direction is the same as the direction of the arrow.

8. Now, we see in the top-right corner of the workflow in *Figure 9.8* that the workflow will run daily at a specified time. We can also write more complex `cron` jobs to schedule the workflow. In that case, the field will take a `cron` expression as a string, such as `schedule_interval="<cron expression>"`. We discussed `cron` expressions in *Chapter 6, Batch Model Serving*. To find out more about scheduling, visit `https://airflow.apache.org/docs/apache-airflow/1.10.1/scheduler.html`. However, we can also manually start the workflow any time we want. Let's click the blue start button, as shown in *Figure 9.9*, and then click the **Trigger DAG** option.

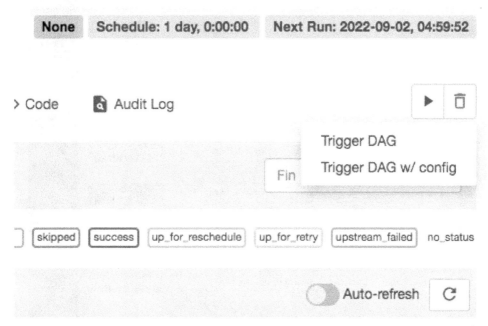

Figure 9.9 – Button to start the workflow manually

After both stages have finished running, we will see that the border of both stages has turned green, as shown in *Figure 9.10*. The legends above the **Auto-refresh** button show the meaning of the different border colors.

Figure 9.10 – All the stages of the pipeline completed

Later stages in the pipeline can only start after the previous stages have finished successfully. You can see this change of color interactively in the UI. For example, in our pipeline, **stage2** will only start after **stage1** is finished. We can confirm this by looking at the log. Let's click on **stage1** of the pipeline, and we will see a pop-up window, as shown in *Figure 9.11*:

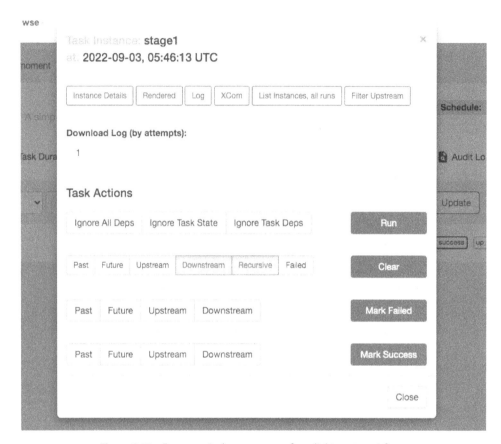

Figure 9.11 – Pop-up window appears after clicking stage1 from which we can see the logs and execution history

By clicking the **Log** button, we can see the **stage1** log and note the following two lines from the log, which prints the following output from the code in the stage1.py Python file that is being run inside **stage1**:

```
[2022-09-03, 06:00:07 UTC] {subprocess.py:92} INFO - I am at
stage 1
[2022-09-03, 06:00:07 UTC] {subprocess.py:92} INFO - After 5
seconds I am existing stage 1
```

Similarly, we check the **stage2** log and note the following line, which prints the output from the code in the `stage2.py` Python file, which is run inside **stage2**:

```
[2022-09-03, 06:00:12 UTC] {subprocess.py:92} INFO - I am at
stage 2
```

From the highlighted times of the logs from the two stages, we can see that the output of **stage2** was generated after the execution of **stage1** completed. So, we have evidence that a later stage in the directional path starts only after the previous stages are complete.

In this section, you were introduced to Apache Airflow and installed it, and saw how to view the pipelines using the Airflow UI, how to run a pipeline, and how to view the output. In the next section, we will create a dummy ML pipeline using Airflow and see how to create a pipeline to serve an ML model following the pipeline pattern.

Demonstrating a machine learning pipeline using Airflow

In this section, we will create a dummy ML pipeline. The pipeline will have the following stages:

- One stage for initializing data and model directories if they are not present
- Two stages for data collection from two different sources
- A stage where we combine the data from the two data collection stages
- A training stage

You can have multiple stages depending on the complexity of your end-to-end process. Let's take a look:

1. First, let's create the DAG using the following code snippet:

```
with DAG(
    'dummy_ml_pipeline',
    description='A dummy machine learning pipeline',
    schedule_interval="0/5 * * * *",
    tags=['ml_pipeline'],
) as dag:
    init_data_directory >> [data_collection_source1,
data_collection_source2] >> data_combiner >> training
```

In the preceding code snippet, we created a pipeline by connecting the tasks or stages using `init_data_directory >> [data_collection_source1, data_collection_source2] >> data_combiner >> training`. Between the first and second `>>` operators, we have passed a list of two stages. This means that the `data_combiner` stage will only start if both the `data_collection_source1` and `data_collection_source2` stages are complete. It will be clearer if we look at the pipeline in the UI, as shown in *Figure 9.12*:

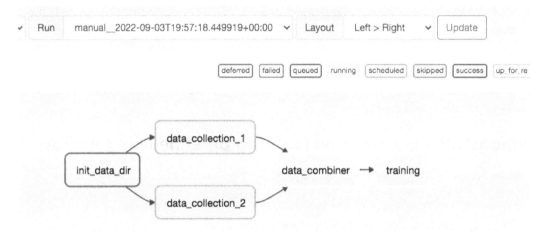

Figure 9.12 – A dummy pipeline with multiple dependencies

In the code snippet, we have omitted some of the parameters from the constructor that are the same as the pipeline we created before. We have changed the `id`, `description`, `schedule_interval`, and `tags` parameters of the DAG. The `schedule_interval` parameter is now given a `"0/5 * * * *"` cron expression. The meaning of this cron expression is that the `cron` job will run every 5 minutes and start the pipeline. This is how we can schedule our pipeline to run automatically at a certain time interval.

2. These stages are created using the following code snippet with `BashOperator`:

    ```
    init_data_directory = BashOperator(
            task_id='init_data_dir',
            bash_command='python3 /Users/mislam/Desktop/
    airflow-examples/dags/stages/init_data_dir.py',
        )


        data_collection_source1 = BashOperator(
            task_id='data_collection_1',
            bash_command='python3 /Users/mislam/Desktop/
    airflow-examples/dags/stages/data_collector_source1.py',
    ```

```
    )

    data_collection_source2 = BashOperator(
        task_id='data_collection_2',
        bash_command='python3 /Users/mislam/Desktop/
airflow-examples/dags/stages/data_collector_source2.py',
    )

    data_combiner = BashOperator(
        task_id='data_combiner',
        bash_command='python3 /Users/mislam/Desktop/
airflow-examples/dags/stages/data_combiner.py',
    )

    training = BashOperator(
        task_id='training',
        bash_command='python3 /Users/mislam/Desktop/
airflow-examples/dags/stages/train.py',
    )
```

> **Note**
> You may notice that in these operators, we are using the path to different files and codes from the machine used to write these codes. You will have to change these paths based on the locations on your own machine. The same suggestion applies to all the examples we are using throughout the chapter.

3. These stages just run different Python files. The Python files contain dummy operations. For example, the init_data_dir.py file contains the following code:

```
import os
if not os.path.exists("/Users/mislam/Desktop/airflow-
examples/dags/stages/data"):
    os.mkdir("/Users/mislam/Desktop/airflow-examples/
dags/stages/data")
    print("The directory data is created")

if not os.path.exists("/Users/mislam/Desktop/airflow-
examples/dags/stages/model_location"):
    os.mkdir("/Users/mislam/Desktop/airflow-examples/
```

```
        dags/stages/model_location")
            print("The directory model_location is created")
```

It creates two directories, data, and model_location, for storing the data and model respectively in later stages.

The data_collector_source1.py file contains the following code snippet:

```
import pandas as pd


df1 = pd.DataFrame({"x": [2, 2, 2, 3, 4, 5], "y": [1, 1,
1, 1, 1, 1]})


df1.to_csv("/Users/mislam/Desktop/airflow-examples/dags/
stages/data/data1.csv")
```

So, we write some dummy data in a CSV file in this code. We do a similar thing in the data_collector_source2.py file. In this file, we combine the data from these two data collection stages and write the combined data to a CSV file. In the training stage, we train a LogisticRegression model and save the trained model to the model_location folder created in the first stage.

4.   We can wait till the pipeline finishes and then check that the data and model_location folders have been created and that the model has been saved in the model_location folder, as shown in *Figure 9.13*:

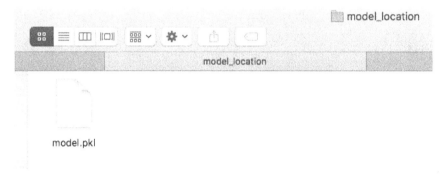

Figure 9.13 – Model is saved successfully after the execution of the pipeline is finished

We have saved this model to our local directory. We can save it inside a server, which will be used by the APIs, and we can create a prediction API to use the model for prediction.

5.   To confirm whether the pipeline is scheduled to run every 5 minutes, let's go to the Airflow UI and look at the scheduling information in the top-right corner. The scheduling information of our pipeline is shown in *Figure 9.14*:

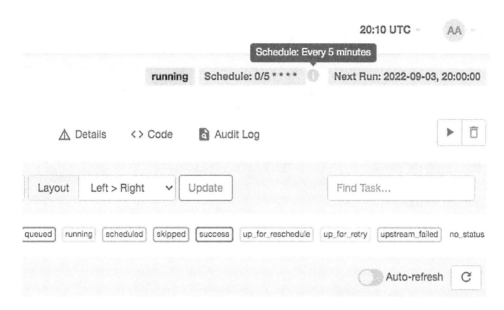

Figure 9.14 – The pipeline is scheduled to run every 5 minutes using a cron expression

In this section, we have seen how to create an ML serving pipeline using Apache Airflow. In the next section, we will see some advantages and disadvantages of the pipeline pattern.

Advantages and disadvantages of the pipeline pattern

In this section, we will explore some of the advantages and disadvantages of the pipeline pattern.

First of all, the advantages of the pipeline pattern are as follows:

- We can split different critical operations of the end-to-end ML processes into the stages of a pipeline. Therefore, if a particular stage fails, we can fix that stage and restart the pipeline from there instead of starting from scratch. If you click on a pipeline stage, you will see that there is an option to start that stage without running the previous stages. You can use the **Run** button shown in *Figure 9.11* to start a pipeline from a particular stage.

- You can monitor the pipeline using a UI and create pipelines with different structures using the operators provided by Apache Airflow.

- You can schedule the pipeline using the options provided by Apache Airflow, and thus, the pipeline can easily be run periodically to get the most up-to-date model.

- You can integrate pipelines with web servers to provide client APIs. You can save the model to a location that is referenced by the web server. In this way, you can avoid uploading the model to the server after the training is complete.

The advantages mentioned here make the pipeline pattern a very good option to choose if the ML application we want to deploy has a lot of stages.

Some of the disadvantages of the pipeline pattern are as follows:

- We need to be careful when setting up dependencies among the stages. For example, let's assume stage *s1* collects data for training and another stage, *s2*, uses the data collected by *s1*. We need to have *s1* before *s2* in the pipeline. If we somehow miss these dependencies, our pipeline will not work.

- Maintaining the pipeline server adds additional technical overhead. A team that needs to use the pipeline pattern might need a dedicated engineer who is an expert on the tool used to create pipelines.

In this section, we have seen some of the advantages and disadvantages of the pipeline pattern. In the next section, we will summarize the chapter and conclude.

Summary

In this chapter, we explored the pipeline model serving pattern and discussed how DAGs can be used to create a pipeline. We have also covered some fundamental concepts on DAG to help you understand what DAGs are.

Then we introduced a tool called Apache Airflow, which can be used to create pipelines. We saw how to get started with Apache Airflow and how to use the operators provided by Apache Airflow to create a pipeline. We saw how dependencies are created using Apache Airflow and how to create separate stages using separate Python files.

We then created a dummy ML pipeline for collecting and combining data, training a model using the data, and then saving the model to a location that is accessible by the server. We explored how to create many-to-one dependencies among the stages for when a stage's actions depend on the completion of multiple stages.

Finally, we discussed the advantages and disadvantages of the pipeline pattern. In the next chapter, we will discuss the ensemble pattern. We will see how multiple models can be combined and served.

Further reading

- You can read about Apache Airflow at `https://airflow.apache.org/docs/`

- You can learn about the Apache Airflow scheduler at `https://airflow.apache.org/docs/apache-airflow/stable/concepts/scheduler.html`

- To find out more about DAGs, you can visit `https://bookdown.org/jbrophy115/bookdown-clinepi/causal.html`

10

Ensemble Model Serving Pattern

In this chapter, we will discuss the ensemble model serving pattern. In the ensemble pattern, we combine the output from multiple models before serving a response to the client. This combination of responses from multiple sources is needed in many scenarios – for example, to get information about audio and video using separate models from a video file, and then combining that information to generate the final inference about the video. We can also combine the output from multiple similar models to make inferences with higher confidence. We will discuss some of these cases in this chapter. We will also explore a dummy end-to-end example of how we can combine multiple models to generate the final response.

At a high level, we are going to cover the following main topics in this chapter:

- Introducing the ensemble pattern
- Using ensemble pattern techniques
- End-to-end dummy example of serving the model

Technical requirements

In this chapter, we will mostly use the same libraries that we have used in previous chapters. You should have Postman or another REST API client installed to be able to send API calls and see the response. All the code for this chapter is provided at this link: `https://github.com/PacktPublishing/Machine-Learning-Model-Serving-Patterns-and-Best-Practices/tree/main/Chapter%2010`.

If you `ModuleNotFoundError` appears while trying to import any library, then you should install the module using the `pip3 install <module_name>` command.

Introducing the ensemble pattern

In this section, we will discuss the ensemble pattern of serving models and the different types of ensembles that can be used to serve a model.

In the ensemble pattern of serving models, more than one model is served together. In an ensemble pattern, an inference decision is made by combining the inferences from all the models in the ensemble.

The final response, Y, from the input, X, will be generated as a combined inference from the models $M1, M2, \ldots Mn$, as shown in the following equation:

$$Y = F(M1(X), M2(X), \ldots Mn(X))$$

In this equation, Y is the response and F is the combination function that combines the responses from all the models. $M1, M2, \ldots Mn$ are different models and X is the input passed to the models.

We can ensemble multiple models for various scenarios. The first four of these different types are introduced in this article: `https://www.anyscale.com/blog/serving-ml-models-in-production-common-patterns`. The following scenarios are examples of where the ensemble pattern can be of great use:

- **Model update**: When updating a model for a sensitive business scenario where a sudden performance drop cannot be afforded, we can use the ensemble pattern. We can run the old model and the new, updated models together for a certain period. During that period, we will keep providing responses to the customer from the old model and use the responses from the new model to verify its performance compared to the old model. When we are confident about the performance of the new model, we can then replace the old model with the new one. In this case, although the final response is still coming from a single model, more than one model is being used and the responses from the other models are utilized to complete the updated process. This method is widely known as **staged rollout** in software engineering. In a staged rollout, the updated software is only made available to a small group of users instead of providing the update for all users instantly.

  To learn more about staged rollouts, please follow this link: `https://developerexperience.io/practices/staged-rollout`. Providing a response to clients from a non-production model instead of a production model is known as traffic shadowing in software engineering.

  To learn more about traffic shadowing, please follow this link: `https://www.getambassador.io/docs/edge-stack/latest/topics/using/shadowing`. This traffic shadowing pattern in a production environment helps to provide nearly zero timeouts of the production application and test the updated model sufficiently before releasing it to production.

- **Aggregation**: To provide responses to users with higher confidence, we can aggregate the responses from similar models. For example, for a classification problem, we can take predictions from different models trained using the same training data and then use the **majority vote** (`https://en.wikipedia.org/wiki/Boyer%E2%80%93Moore_majority_vote_algorithm`) algorithm to find the majority class predicted by the models. Then, we can provide that majority class as the final prediction response to the client. For regression problems, instead of majority voting, we can take an average of all the predictions from different models and then provide the response to the clients.

- **Model selection**: We can serve multiple models and, based on some feature in the input, select a particular prediction model. For example, let's say we have models for predicting the class of different types of objects, such as flowers or animals. We can have separate models for each of these objects and, based on the input, we can select the appropriate model.

- **Combining different responses**: For tasks such as describing objects, we can use multiple models to describe different aspects of the object. For example, let's say we are describing different features of a house, such as its height and color. We can use multiple models for each of the subanalysis tasks and then combine the responses to provide a full description.

- **Serving degraded response:** In this kind of ensemble, usually, two models are served in parallel. One model is strong and can provide a more accurate response but can time out because the model usually has time-consuming computations. Another model is served in parallel, which is simple but can provide a response very quickly – although it may be less trustworthy. If a client request comes, we send the request to both models. If the robust model times out, we send the degraded response from the weaker model. For example, let's say we have a system for language translation. We might serve a strong model in parallel with a weak model. If the strong model times out when handling a client request, we are able to provide a degraded response from the weaker model.

In this section, we have introduced you to the ensemble pattern and discussed different kinds of ensemble approaches that can be used. In the next section, we will discuss these approaches in the ensemble pattern along with examples.

Using ensemble pattern techniques

In this section, we will discuss different types of ensemble approaches along with examples. We have seen that we can combine the models in five types of different scenarios. The following subsections will discuss them one by one.

Model update

In the **machine learning** (**ML**) deployment life cycle, updating the model happens regularly. For example, we might have to update a model for route planning if new roads and infrastructure are built or removed. Whenever a model needs to be replaced, it might be risky to replace the current model directly. If for some reason, the new model performs poorly compared to the previous model, then it might cause critical business problems and loss of trust. For example, let's imagine we have updated a model with a new version tag, V2, that predicts a stock price. The V1 model version was predicting stock prices with an MSE of 10.0. Although during training the V2 model was performing very well, in production, we noticed that the V2 model was giving an MSE of 20.0. Therefore, if we directly deployed model V2 in replacement of V1, we could lose customer trust. By following the ensemble process of updating the model, we can avoid this risk.

Therefore, in this case, we keep both the old model and the new model in the production system. The models perform the following operations for a certain period that we can call the **evaluation period**:

- **Old model**: Keeps providing predictions as before
- **New model**: Predictions from the new model are compared with the performance of the old model

The results from the new and old models might match in most cases. However, for some inputs, the response might not match. In that case, we have to manually verify which model gave the correct output and then, finally, compute which model showed better accuracy in the differing responses. For example, let's look at the dummy data shown in *Figure 10.1*:

Actual level	Prediction by old model	Prediction by new model
Class A	Class A	Class A
Class B	Class B	Class B
Class A	Class A	Class A
Class B	Class B	Class B
Class A	Class A	Class A
Class B	Class B	Class B
Class A	Class B	Class A
Class A	Class B	Class A
Class A	Class B	Class A
Class B	Class B	Class A

Figure 10.1 – Dummy data showing actual levels and predictions from the old model and the new model

In the table in *Figure 10.1*, the first column shows the actual level, the second column shows the predictions made by the old model, and the third column shows the predictions made by the new model during an evaluation period. The bold rows in the table are the rows where the predictions from the old model and the new model are different. We can see that the old model only made one correct prediction in these four rows and the new model made three correct predictions, so, the prediction accuracy of the new model is ¾ = 75%, and the prediction accuracy of the old model is ¼ = 25%. Therefore, we can decide to use the new model, as the accuracy is satisfactory compared to the old model.

The following code snippet shows a dummy example of using two models as an ensemble while updating:

```
def load_model(filename):
```

```
        print("Loading the model:", filename)

def predict_model_current(X):
    model = load_model("model_update/model_current/model.txt")
    print("Current model is predicting for ", X)
    return "dummy_pred_current"

def predict_model_new(X):
    model = load_model("model_update/model_new/model.txt")
    print("New model is predicting for ", X)
    return "dummy_pred_new"

def predict(evaluation_period, X):
    if evaluation_period == True:
        pred_current = predict_model_current(X)
        pred_new = predict_model_new(X)
        file = open("evaluation_data.csv", "a")
        file.write(f"{pred_current}, {pred_new}")
        return pred_current
    else:
        return predict_model_current(X)
```

In our code, we have not used an actual model; rather, we have stored the dummy text file in the folders meant to save the current and new models. The directory structure of the storage is shown in *Figure 10.2*. We can see from the figure that we have stored just two text files. You will store two different models here after the training:

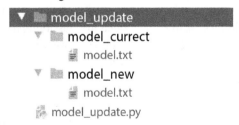

Figure 10.2 – Directory structure of storing two parallel models
to use as an ensemble during model update

As shown in the code snippet, during the evaluation period, we will get a prediction from both of these models. However, we only return `pred_current` from the current model to the user. We save both of the predictions to a separate file that will be used for evaluation later on. We write the predictions that we got from the last code snippet to a CSV file using the following code:

```
file = open("evaluation_data.csv", "a")
file.write(f"{pred_current}, {pred_new}")
```

Then, we can manually move the new model to the current model directory if the evaluation results succeed. Note that in the `predict` method, we use both models. The first model is used if the condition of the `if` statement is `true`; otherwise, we use the other model inside the `else` block as seen from the `predict` method code snippet. Predictions from the current model are accessed using the `predict_model_current` API and predictions from the new model are accessed using the `predict_model_new` API.

This is how ensembles work during a model update. Although during the actual deployment, there will be actual models instead of dummy files and the locations can also be two totally different servers, the concept of model update using the ensemble pattern will remain the same.

Aggregation

In this approach, we aggregate the response from multiple models and send the aggregated response to the users. Aggregation usually happens in the following two ways for regression and classification problems, respectively:

- For regression problems, we take the average of the response from multiple models and use the average as the prediction. For example, let's say we have two models to predict the price of a house. One is M1, a `RandomForestRegression` model, and the other is M2, an `AdaBoostRegression` model. Let's assume M1 has predicted the price of a house as $100,000 and M2 has predicted the price of the house as $120,000. We aggregate the response from them by taking the average and returning ($100,000 + $120,000) / 2 = $110,000 to the clients.

- For classification problems, we take the majority class that is selected by the models as the prediction. For example, let's assume five models are classifying handwritten digits. The five models respectively predict [1, 2, 1, 2, 1]. Here, we will take the majority class that has been predicted. We notice that 1 has been predicted by three models and 2 has been predicted by two models. Therefore, we will return 1 as the final predicted class to the client.

Let's assume that we have three models, M1, M2, and M3, served in the ensemble pattern to predict the price of a stock. Let's say the feature set of the stock is X and we need to predict the price with this feature set. The responses from the models are as follows:

```
Y1 = M1(X)
Y2 = M2(X)
Y3 = M3(X)
```

Therefore, the final response that will be returned to the user is the following:

```
Y = (Y1 + Y2 + Y3)/3
```

Usually, averaging the responses helps provide a more accurate prediction compared to the prediction from a single model. Therefore, the response from the ensemble pattern creates more confidence among the customers.

For example, let's say the actual price of the stock is $10. The M1, M2, and M3 models made predictions of $8, $12, and $13, respectively.

The average of these three predictions is `(8 + 12 + 13)/3 = 11`.

We notice that although the predictions made by the models differ by -$2, $2, and $3, respectively, the aggregated response differs by $1, so we got a response that is closer to the actual price compared to the predictions made by the individual models. It's not necessary for the aggregate response to always perform better than all the individual models, but most of the time, the average prediction will outperform the individual predictions. For example, if the three models make predictions of $11, $10, and $11, respectively, the average response is ~$10.67. Although this is better than the prediction from the first and third models, it is not better than the prediction from the second model.

Besides the aggregation techniques we just defined, there can be many other aggregation techniques for both regression and classification tasks. For example, instead of taking the direct mean, we can use the geometric mean for aggregating the results of multiple models, we can use weighted means to provide different weights to different models, and so on. For classification, we can also use different variants of majority class selection algorithms, weighted majority algorithms, and so on.

In the next sub-section, we will discuss selecting the majority class in the classification problem using the Boyer-Moore algorithm.

Selecting the majority class

For classification problems, we can use the majority voting algorithm to select the class. For example, let's say we have three classification models: M1, M2, and M3. The models make predictions, as follows, for an input, X:

```
C1 = M1(X)
C1 = M2(X)
C2 = M3(X)
```

Now, we can use the majority voting algorithm to find out which class has been predicted by the highest number of models. In the preceding example, the majority class is C1, which is predicted by two out of the three models. One version of the majority algorithm is known as the **Boyer-Moore majority voting algorithm**, which finds an element that appears at least N/2 times out of the N numbers in a list or sequence.

For example, if the input array is [1 1 1 3], the Boyer-Moore algorithm will say the majority element is 1, as it appeared more than 4/2 = 2 times. Here, N = 4 is the number of elements in the array.

> **Boyer-Moore**
>
> To learn more about the Boyer-Moore algorithm, you can check out the following link: https://www.geeksforgeeks.org/boyer-moore-majority-voting-algorithm/.

In our aggregation strategy, the Boyer-Moore approach will not always work. For example, let's assume we have six models and they make predictions of [C1 C1 C2 C3 C4 C5]. Here, none of the elements appeared more than 6/2 = 3 times. We still need to provide the prediction as C1, as it appeared the highest number of times. Therefore, we can use a majority voting algorithm that will select the class that has been predicted by the highest number of models. An example of finding the majority class is as follows:

```
from collections import Counter

x = ['C1', 'C1', 'C1', 'C2', 'C2', 'C3', 'C3']
counts = Counter(x)
print("Counts of different elements", counts)
major_element = counts.most_common(1)[0][0]
print("Major element", major_element)
```

The output of the preceding program is as follows:

```
Counts of different elements Counter({'C1': 3, 'C2': 2, 'C3': 2})
Major element C1
```

In this program, first, we compute the count of different classes in the array and then select the class that has the highest count using the major_element = counts.most_common(1)[0][0] line. Here, 1 is passed as an argument to the most_common(n) method to select the top most common element. This method returns an array of tuples, so we select the first element of the first tuple in the array using the [0][0] indices.

Model selection

We can serve different models that have been specialized for different problems following the ensemble pattern. For example, we can have a model to detect the names of different fruits. Different fruits have different features and therefore the prediction task of the fruits can be seen as separate problems; we can have separate models to solve each of those problems. We can serve these models together to form a complete fruit detection system.

Based on a feature in the input, we will select the appropriate model. For example, let's say we want to design an ML system that can detect the class of a fruit and the class of a flower. The features for flower detection and fruit detection will be different. There can be two different models for handling the inference for these two different inputs: one for flowers and the other for fruit. An example of this is shown in *Figure 10.3*, where we select the model to detect either a flower or fruit based on the input:

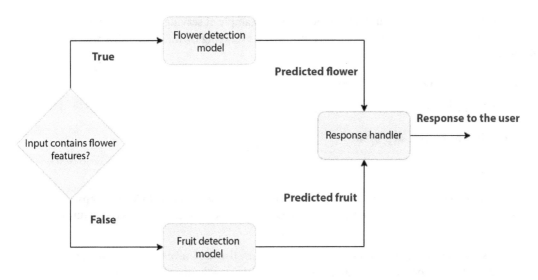

Figure 10.3 – Ensemble pattern serving two models with the option to select a particular model

We can aggregate the responses from more than one model for flowers, and more than one model for fruits, which makes serving more complicated. Therefore, instead of a single model for a particular problem, we will have multiple models. The responses will be aggregated before providing an overall response to the client.

Combining responses

ML is acquiring more and more responsibilities as time goes on. We now use ML models to describe pictures and contexts, drive autonomous vehicles, and so on. In many of these scenarios, we might need the responses from multiple models to be combined. For example, let's say we are describing a painting and we want the following descriptions:

- Detect what colors are used in the picture

- Find out the painter's name from the signature

- Detect different objects in the picture

We might create a separate model for each of these cases and then combine the responses to provide an overall summary of the painting.

In this section, we have seen different techniques in the ensemble pattern. In the next section, we will discuss an end-to-end example of serving two regression models together, and then we will conclude the chapter.

End-to-end dummy example of serving the model

In this section, we will create an end-to-end dummy example of serving the two regression models together, and then we will combine their responses by averaging them. The models we will use are the following:

- The RandomForestRegression model

- The AdaBoostRegression model

Let's describe the process step-by-step:

> **Note**
>
> Please keep in mind that the output may be different in your case from the following steps, as you will train the model with some random data generated using the make_regression function.

1. First, let's create the two models with some dummy data and save the models using pickle. The following code snippet creates the models and saves the trained models:

    ```
    from sklearn.ensemble import RandomForestRegressor,
    AdaBoostRegressor
    from sklearn.datasets import make_regression
    import pickle
    X, y = make_regression(n_features=2, random_state=0,
    shuffle=False, n_samples=20)
    model1 = RandomForestRegressor(max_depth=2)
    model1.fit(X, y)
    print(model1.predict([[0, 0]]))
    pickle.dump(model1, open("rfreg.pb", "wb"))
    model2 = AdaBoostRegressor(n_estimators=5)
    ```

```
model2.fit(X, y)
pickle.dump(model2, open("adaboostreg.pb", "wb"))
```

2. Now, we create another file that will be used to handle client requests. We will not create a server using Flask, as we have already shown how to do so in earlier chapters. Feel free to make a Flask API for this step. We will use the following code snippet to create a dummy serving:

```
from sklearn.ensemble import RandomForestRegressor,
AdaBoostRegressor
import pickle
model1: RandomForestRegressor = pickle.load(open("rfreg.
pb", "rb"))
model2: AdaBoostRegressor = pickle.
load(open("adaboostreg.pb", "rb"))

def predict_model1(X):
    response = model1.predict(X)
    print("Response from model 1 is", response)
    return response

def predict_model2(X):
    response = model2.predict(X)
    print("Response from model 2 is", response)
    return response
def predict(X):
    response1 = predict_model1(X)
    response2 = predict_model2(X)
    response = (response1 + response2)/2
    print("Final response is ", response)

predict([[0, 0]])
```

In this code, we have two methods to get predictions from two different models. Then, we have the predict method, which will be called by the client through an API. We combine the responses from the two models inside this method and print that final response. If we call the predict method with [[0, 0]], we will get the following output in the console:

```
Response from model 1 is [8.1921427]
```

```
Response from model 2 is [-11.31397077]
Final response is   [-1.56091404]
```

Have a look; we get responses from both models and then average the responses to provide the final response.

In this section, we have discussed a dummy example to demonstrate how can we serve the models in the ensemble pattern. In the next section, we will summarize the chapter and conclude.

Summary

In this chapter, we have discussed the ensemble pattern of model serving. We were introduced to the ensemble pattern and different types of approaches to using it.

We have discussed how this pattern can be of use when we need to carefully update a new model, when we need predictions from multiple models to increase the prediction accuracy, when we need an option for multiple models based on different inputs, and when we need to combine the responses from multiple models to produce a final output.

In the next chapter, we will discuss the business logic pattern to serve ML models. We will discuss how while serving an ML model, we might need different business logic, such as user authentication or querying a database.

11

Business Logic Pattern

In this chapter, we will discuss the **business logic pattern** of serving models. In this pattern, we add some business logic, along with the model inference code. This is essential for successfully serving models because model inference code alone can't meet the client's requirements. We need additional business logic, such as user authentication, data validation, and feature transformation.

At a high level, we are going to cover the following main topics in this chapter:

- Introducing the business logic pattern
- Technical approaches to business logic in model serving

Technical requirements

In this chapter, we will use the same libraries that we used in previous chapters. You should have Postman or another REST API client installed to be able to send API calls and see the response. All the code for this chapter is provided at this link: `https://github.com/PacktPublishing/Machine-Learning-Model-Serving-Patterns-and-Best-Practices/tree/main/Chapter%2011`.

If you `ModuleNotFoundError` appears while trying to import a library, then you should install that module using the `pip3 install <module_name>` command.

Introducing the business logic pattern

In this section, we will introduce you to the business logic pattern of serving models.

When we bring our model to production, some business logic will be required. Business logic is any code that is not directly related to inference by the ML model. We have used some business logic throughout the book for various purposes, such as checking the threshold for updating models and checking the input type. Some examples of business logic are as follows:

- **Authenticating the user**: This kind of business logic is used to check whether a user has permission to call the APIs. We can't keep our APIs public in most cases because this might be very risky, as our critical information can be compromised. If the models are involved in critical business decisions, then the APIs may be restricted to only a few groups of people as well. We need to check the user credentials at the very beginning of the API call. The code snippet that performs authentication in this way is a critical part of model inference code.

- **Determining which model to access**: Sometimes, we will have multiple models. We have to determine which model to access from the server side. The choice of model can be based on the user's role. For example, someone from HR may want to access a model to get the current happiness index of the employees, or an engineering manager may be interested in accessing a model that will provide the performance index of the employees. Based on the user role, we can redirect to different models for inference. The choice of the model may be based on the type of input. Suppose there are different models for detecting flowers and fruits. Based on the input features, we have to determine which model to use during inference. This logical operation becomes critical when multiple models are served together in an ensemble model serving approach.

- **Data validation**: The validation of input data is important in model serving to reduce the number of inference errors. If we have data validation, we can throw an appropriate error to the customer, indicating that this is a client-side error and our model is still in good health. If we do not have this data validation, then due to the wrong data passed by clients, the errors might give a false impression that the model is bad. This will affect the reputation of the engineering team and will result in poor client satisfaction. Therefore, we should have different data validations, such as the dimensions of the data, the type of data, and the normalization of the data. All these validations ensure some properties of data. That's why these validations are placed within the category of data validation.

- **Writing logs to server**: We have to write logs from our server code. This is essential for auditing user access, errors, warnings, data access patterns, and so on, and in understanding how our system behaves over different calls. We can also use logs to create operational metrics such as failure rate and latency. Writing logs can happen before the inference as well as after the inference. For example, let's say we want to monitor the time taken by an inference call. We can write a log, `"INFO: Starting the inference at {datetime.now()}"`, and after the call, we can have another log, `"INFO: Finished the inference at {datetime.now}"`. From these two logs, we can determine how much time was needed for inference. Without these logs, we do not know how our served model is performing, how many times the model is being accessed, or who is accessing our model.

- **Database lookup for pre-computed information**: In ML, we might use a lot of pre-computed information. For example, if we need to retrain a model, we can load the features from a database instead of recreating them. If there is some new data, we can recompute the features for that new data; otherwise, we can load the features from the database. We need to add this logic before the model inference.

- **Feature transformation**: If we have raw data, we need to transform the raw data into features. For example, let's say the input has a Boolean feature with [True, False] values. We might have to encode the feature to [0, 1]. This logic is essential not only during training but also during inference because, during inference, the user might pass raw data.

- **Sending notifications**: Sending notifications to different stakeholders after the inference can also be necessary in some cases. We need to add this business logic after the model inference is complete. The notifications might be sent via email, simple notification service, text message, or using various other methods.

We have learned about some examples of business logic under different categories. Next, we will how we can divide a lot of the business logic into two types of business logic.

Type of business logic

In the previous section, we looked at different types of business logic. However, based on the location of the business logic code in the inference file, we can divide the business logic in the ML model into two types:

- **Pre-inference business logic**: This business logic needs to be written before the call to the model is made, and this business logic can also work as a guard to prevent unwanted calls to the model. Examples include business logic to authenticate the user, validate the data, and select the right model.

- **Post-inference business logic**: This business logic needs to be implemented after the model inference is complete. Examples include business logic to write the inference results to a database, mapping the inference classes to appropriate labels, and sending notifications.

The location of these types can be seen in *Figure 11.1*. We can see that the model is called only if the pre-inference logic passes. Otherwise, an error is shown to the caller. After the model is invoked, we have the post-inference logic.

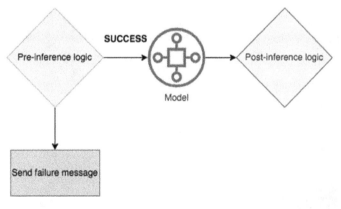

Figure 11.1 – Location of the pre-inference and post-inference business logic

In this section, we have seen some of the business logic that is used as part of serving an ML model. We have also seen that business logic can come before the model inference as well as after the model inference. In the next section, we will see technical approaches to some business logic, along with examples.

Technical approaches to business logic in model serving

In the last section, we have learned about the different types of business logic along with examples. Some business logic is exclusive to ML serving, such as data validation, feature transformation, and so on. Some business logic is common to any kind of application – for example, user authentication, writing logs, and accessing databases. We will explain some business logic that is crucial to ML in the following sub-sections.

Data validation

Data validation is very important in serving ML models. In most of the models in our previous chapters, we have assumed the user will pass data in the right format, but that may not always be the case. For example, let's say the model needs input in the format of [[int, int]] to make inferences, and if the user does not follow the format, we will get errors.

For example, let's consider the following code snippet:

```
model = RandomForestRegressor(max_depth=2)
model.fit(X, y)
print(model.predict([0, 0]))
```

Here, we are trying to pass [0, 0] to the model for prediction. However, we get the following error:

```
raise ValueError(
  ValueError: Expected 2D array, got 1D array instead:
    array=[0. 0.].
```

Therefore, we can add validation business logic before passing the input to the model that will check whether the shape of the input is correct.

We can add the following business logic before passing the input to the model for prediction:

```
Xt = np.array([0, 0])
if len(Xt.shape) != 2:
    print(f"Shape {Xt.shape} is not correct ")
```

```
else:
    print(model.predict(Xt))
```

We are passing Xt data of the wrong shape in the preceding code snippet, so we get the following output:

Shape (2,) is not correct

We previously got an error that stopped our program. Now, as we are validating the input, we can avoid that exception. We can also raise a 4XX error from this if block by telling the client that the error is a client-side error with a readable error message.

We can also check the type of data that is passed as input to the model before the model is called. For example, let's look at the following inference, where the input is a string that can't be converted into a floating-point number:

```
Xt = np.array([["Zero", "Zero"]])
print(model.predict(Xt))
```

Therefore, we get the following error from the code snippet:

```
ValueError: could not convert string to float: 'Zero'
```

To solve this error, we can add the following validation code:

```
Xt = np.array([["Zero", "Zero"]])
try:
    Xt = Xt.astype(np.float64)
    print("Floating point data", Xt)
    print(model.predict(Xt))
except:
    print("Data type is not correct!")
```

The preceding code snippet will throw the following output:

Data type is not correct!

However, if you pass the Xt = np.array([["0", "0"]]) input, you will get the following output:

Floating point data [[0. 0.]]
[6.941918]

So, the 0 string could be converted into a floating point. We can add many other validations such as this. We have to think about the kinds of validation that are needed for our model. The validations may be different for different models. The data validation happens before inference and that's why this logic can be called pre-inference logic.

Feature transformation

Feature transformation is very important for an ML model. Converting the raw data into some suitable features is very important if we want a good ML model. This transformation is also needed during inference. For example, suppose a feature in the input data is `climate` and can contain values of `["Sunny", "Cloudy", "Rainy"]`. Now, the question is how we can pass this information to an ML model. The ML model can only deal with numeric data. The solution to this is one-hot encoding. For example, see the following code snippet:

```
import pandas as pd

df = pd.DataFrame({"climate": ["Sunny","Rainy","Cloudy"]})
print("Initial data")
print(df.head())
df2 = pd.get_dummies(df)
print("Data after one hot encoding")
print(df2)
```

In this code snippet, we are encoding the data that will convert the categorical data for the `climate` feature into numerical data. The output from the preceding code snippet is as follows:

```
Initial data
  climate
0  Sunny
1  Rainy
2 Cloudy
Data after one hot encoding
   climate_Cloudy  climate_Rainy  climate_Sunny
0               0              0              1
1               0              1              0
2               1              0              0
```

After encoding, three features are created from the single `climate` feature. The new features are `climate_Cloudy`, `climate_Rainy`, and `climate_Sunny`. The value for the particular feature is 0 in the row for which the feature is present. For example, for the first row at index 0, the climate is Sunny, that's why the value of `climate_Sunny` for the first row is 1.

The feature transformation logic needs to come before the model inference or training. That's why this business logic is a pre-inference logic.

Prediction post-processing

Sometimes, the predictions from the model are in a raw format that can't be understood by the client. For example, let's say we have a model predicting the names of flowers. The dataset that is used to train the model has four flowers:

- Rose
- Sunflower
- Marigold
- Lotus

The model can only work with numeric data. Therefore, the model can't provide these names directly during prediction. After the predictions or inferences are done, we need a post-processing logic to map the predictions to the actual names. For example, the model detecting the four flowers can simply provide one of the following predictions:

- 0: The output prediction for roses will be 0
- 1: If the model identifies the flower as a sunflower, then it will predict 1
- 2: If the flower is a marigold, then the model will predict 2
- 3: If the flower is a lotus, then the model will predict 3

Now, suppose we got a batch prediction from the model as follows:

```
[0, 0, 0, 1, 0, 1, 0, 2, 3]
```

We need to map this prediction to the following:

```
['Rose', 'Rose', 'Rose', 'Sunflower', 'Rose', 'Sunflower',
'Marigold', 'Lotus']
```

To do this mapping, we will add business logic such as the following:

```
response = [0, 0, 0, 1, 0, 1, 0, 2, 3]
mapping = {0: "Rose", 1: "Sunflower", 2: "Marigold", 3:
"Lotus"}
converter = lambda x : mapping[x]
final_response = [converter(x) for x in response]
print(final_response)
```

We get the following output from the preceding code snippet:

```
['Rose', 'Rose', 'Rose', 'Sunflower', 'Rose', 'Sunflower',
'Rose', 'Marigold', 'Lotus']
```

So, we can see that `final_response` is a user-friendly response that can be returned to the client. As this business logic is used after the inference, we call it post-inference logic.

In this section, we have discussed some business logic with code examples that are critical for ML model serving and can be found in almost all serving cases.

Summary

In this chapter, we discussed the business logic pattern of serving ML models. We saw how different business logic can be added as preconditions before calling a model. We discussed different kinds of business logic that are essential for serving ML models. With this chapter, we have concluded our discussions of all the patterns of model serving that we wanted to cover in this book.

In the next few chapters, we will discuss some tools for serving ML models, starting with TensorFlow Serving.

Part 3: Introduction to Tools for Model Serving

In this part, we explore some tools for serving machine learning models, and we also demonstrate how to serve machine learning models using these tools.

This part contains the following chapters:

- *Chapter 12, Exploring TensorFlow Serving*
- *Chapter 13, Using Ray Serve*
- *Chapter 14, Using BentoML*

12

Exploring TensorFlow Serving

In this chapter, we will start our discussion by looking at a few model-serving tools. We will cover only a few of the many available tools over the next few chapters. In this chapter, we will talk about **TensorFlow Serving**. TensorFlow Serving is a high-performance tool for serving machine learning models. Though it is mainly used for serving TensorFlow models, it can easily be extended to other kinds of models.

At a high level, we are going to cover the following main topics in this chapter:

- Introducing TensorFlow Serving
- Using TensorFlow Serving to serve models

Technical requirements

In this chapter, we will mostly use the same libraries that we have used in the previous chapters. You should have Postman or another REST API client installed to be able to send API calls and see the responses. All the code for this chapter is provided at `https://github.com/PacktPublishing/Machine-Learning-Model-Serving-Patterns-and-Best-Practices/tree/main/Chapter%2012`.

If you get `ModuleNotFoundError` while trying to import any library, then you should install that module using the `pip3 install <module_name>` command.

You also need to install **Docker** for this chapter. Please make sure that you install Docker from `https://www.docker.com/`. Docker is the easiest and most recommended system for using TensorFlow Serving, as per the official documentation at `https://www.tensorflow.org/tfx/serving/docker`.

Introducing TensorFlow Serving

In this section, we will provide a high-level introduction to the TensorFlow architecture and its key concepts. TensorFlow is designed to provide a high-performance production serving environment for serving machine learning models. TensorFlow provides default integration with TensorFlow models but it can be extended to other models as well, such as scikit-learn. To learn more about integrating other libraries with TensorFlow Serving, please go to `https://www.tensorflow.org/tfx/guide/non_tf`. TensorFlow provides support for easily deploying new models by keeping the architecture and the APIs the same, making provisioning the versioning support of the models in production easier.

To understand the architecture of TensorFlow, you first need to understand the following key concepts.

Servable

A **servable** is the name of the abstraction object that is used by the client for computation or inference. For example, if a client makes a prediction request, the request will go to a servable to provide the prediction. TensorFlow supports versioning a servable. Therefore, whenever we want to update a model in TensorFlow Serving, this concept of versioning the servable comes into the picture.

The following are some examples of a TensorFlow servable:

- A TensorFlow `SavedModelBundle` that represents a saved machine learning model
- A lookup table for embedding or other vocabulary and categories, and so on

In short, a servable is an object or computation unit that's exposed or served to the clients to serve the requests. For example, let's look at the architecture in *Figure 12.1*. The client is interacting with a servable via `ServableHandle` to make requests:

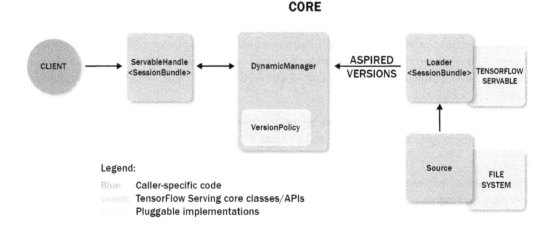

Figure 12.1 – TensorFlow Serving core architecture (source: `https://www.tensorflow.org/tfx/serving/architecture`)

A model in TensorFlow Serving is created as one or more servables. For example, there can be multiple machine learning algorithms in Serving that will require multiple servables. There can also be lookup tables within a servable that can be used for feature mapping, getting encodings, embeddings, and so on.

Loader

Loaders are used to load and unload a servable.

In *Figure 12.1*, we can see that a **Loader** loads the **TENSORFLOW SERVABLE** and sends the **ASPIRED VERSIONS** of the servable to `Manager`, which handles the life cycle of a servable. We will discuss the aspired version later.

Source

A source creates loaders for servable versions. The source creates one or more servable streams. And for each stream, the source provides a **Loader**. A servable stream is the sequence of versions for a servable that are loaded from the filesystem, as shown in *Figure 12.1*. The servable stream presents a sequence of all the servables sorted by version number. The latest version will be used for serving by the **Manager** by default. The servable sequence provides a stream of all versions so that you can choose a different version of the servable than the latest version.

Aspired versions

This is the set of versions of the servables that need to be loaded and made ready. The source block provides the set of aspired versions for a servable stream at a time. When a new set of aspired versions are given to the **DynamicManager** block, as shown in *Figure 12.1*, the old aspired versions are replaced by the new ones. This is how versioning is handled in TensorFlow Serving.

Manager

A **Manager**, also known as **DynamicManager**, as shown in *Figure 12.1*, handles the life cycle of servables. The **Manager** receives the aspired versions from the source through loaders and then serves the new servables, which will be exposed to the clients, and it also unloads the previously served servables. The **Manager** contains the version policy.

The version policy can be of the following two types:

- **Availability-preserving policy**: This policy does not tolerate periods of unavailability between loading a new version and unloading an old version. So, before unloading the old version, the new version is loaded in parallel, and then the old version can be safely unloaded without any downtime.
- **Resource-preserving policy**: This policy is more concerned about resource frugality. It does not support the resources needed to serve two versions in parallel. So, before loading the new version, this policy requires the old version to be unloaded.

In summary, the different blocks do the following operations during the life cycle of serving a model using TensorFlow Serving:

1. The **Source** creates a **Loader** for a specific version of the servable. The **Loader** will contain all the necessary metadata needed to load the servable. For example, if the servable is a lookup table that reads data from a CSV file in the filesystem, the source will create a **Loader** for a specific version of the file. The source will detect whether a new version of the file is available.

2. The source then notifies the **Manager** that a new version of the data is available.

3. The **Manager** has a version policy, as shown in *Figure 12.1*. Using the version policy, the **Manager** decides how to unload the old version and load the new version.

4. The **Manager** then asks the Loader to load the new servable. The Loader then creates an instance of a lookup table from the CSV source.

5. The client sends a request for the servable. The client can specify the version or just ask for the default latest version of the servable. The **Manager** returns a handle for the servable.

An example to help you understand these steps is provided on the official TensorFlow website at `https://www.tensorflow.org/tfx/serving/architecture#life_of_a_servable`.

In this section, we introduced TensorFlow Serving and have seen a high-level overview of the components in the core architecture. In the next section, we will discuss serving models using TensorFlow Serving. To show the end-to-end process, we will use the demo models already provided by TensorFlow Serving in its official repository to save us the effort of building models ourselves; building a model is not the goal of our chapter. We only want to show how the model can be served. The examples will mainly focus on educating you on the steps to use TensorFlow Serving so that you can use these steps to serve your own models.

Using TensorFlow Serving to serve models

In this section, we will use TensorFlow Serving to serve models. First, we will use the recommended mechanism of using TensorFlow with Docker. The official page presenting this example is `https://www.tensorflow.org/tfx/serving/docker`.

TensorFlow Serving with Docker

Make sure Docker is installed on your platform. Follow the link provided in the *Technical requirements* section to start Docker. Now, let's work through the following steps to serve our dummy example model:

1. First of all, start Docker and make sure it is running. You can verify whether Docker is running by running the `docker -version` command in your operating system's terminal. It should give you an output similar to the following:

    ```
    →  TF_SERVE docker --version
    Docker version 20.10.11, build dea9396
    ```

2. Now, let's pull the latest TensorFlow Docker image using the `docker pull tensorflow/serving` command. You should see the following output:

```
→  TF_SERVE docker pull tensorflow/serving
Using default tag: latest
latest: Pulling from tensorflow/serving
726b8a513d66: Pull complete
ebf1b93c73ad: Pull complete
9683aef6d08b: Pull complete
fdb82a459cb8: Pull complete
a1e1d413e326: Pull complete
Digest: sha256:6c3c199683df6165f5ae28266131722063e9fa012c1
5065fc4e245ac7d1db980
Status: Downloaded newer image for tensorflow/
serving:latest
docker.io/tensorflow/serving:latest
```

3. Next, let's clone the GitHub repository for TensorFlow Serving using the `git clone https://github.com/tensorflow/serving` command. This contains a bunch of demo models. We will use one of these demo models to demonstrate serving. You should see the repository cloned in your current directory. Navigate through the repository to get familiar with it. We will use the `saved_model_half_plus_two_cpu` model from this folder, as shown in *Figure 12.2*, to demonstrate the serving process. We chose this model because it is a simple mathematical model and we can easily analyze the input and the predicted result to better understand the end-to-end serving process. This model takes an input, halves it, and then adds two to the result. The final result is returned to the client.

4. Let's create an environment variable in the terminal with the path to the example demo models by running the following command:

```
TESTDATA="$(pwd)/serving/tensorflow_serving/servables/
tensorflow/testdata"
```

This test data directory contains a bunch of servables that can be served directly without us having to create our own servable from scratch. If you look at the directory, you will see we have the servables shown in *Figure 12.2*. These servables can be directly loaded for serving.

```
→  TF_SERVE cd serving/tensorflow_serving/servables/tensorflow/testdata/
→  testdata git:(master) ls
BUILD                                   monitoring_config.txt                                  saved_model_half_plus_two_gpu
bad_half_plus_two                       parse_example_tflite                                   saved_model_half_plus_two_gpu_trt
bad_model_config.txt                    parse_example_tflite.README                            saved_model_half_plus_two_mkl
batching_config.txt                     parse_example_tflite.py                                saved_model_half_plus_two_mlmd
export_bad_half_plus_two.py             saved_model_counter                                    saved_model_half_plus_two_tf2_2_versions_metadata.json
export_counter.py                       saved_model_half_plus_three                            saved_model_half_plus_two_tf2_cpu
good_model_config.txt                   saved_model_half_plus_two.py                           saved_model_half_plus_two_tflite
half_plus_two_model_metadata.json       saved_model_half_plus_two_2_versions                   saved_model_half_plus_two_tflite_with_sigdef
mobilenet_v1_quant_tflite               saved_model_half_plus_two_2_versions_metadata.json     tf_text_regression
mobilenet_v1_quant_tflite.README        saved_model_half_plus_two_cpu                          tf_text_regression.README
```

Figure 12.2 – Demo servables provided by the TensorFlow Serving repository

5. We will be using the `saved_model_half_plus_two_cpu` servable for our example. Let's go inside the repository to see what we have. We have another directory, as shown in *Figure 12.3*:

```
→  testdata git:(master) cd saved_model_half_plus_two_cpu
→  saved_model_half_plus_two_cpu git:(master) ls
00000123
→  saved_model_half_plus_two_cpu git:(master)
```

Figure 12.3 – The servable directory contains a subdirectory for different versions

This `0000123` subdirectory is usually used to denote different versions. Let's go inside this version directory and see what we have there.

6. It contains a saved TensorFlow model in `.pb` format, as shown in *Figure 12.4*. It also contains a directory for variables and another directory for assets. This is the ideal structure that we get after saving a TensorFlow model:

```
→  saved_model_half_plus_two_cpu git:(master) cd 00000123
→  00000123 git:(master) ls
assets           saved_model.pb variables
→  00000123 git:(master)
```

Figure 12.4 – The 0000123 version folder contains a saved TensorFlow model

7. Now, we can serve the model using the following command:

```
docker run -t --rm -p 8501:8501 \
    -v "$TESTDATA/saved_model_half_plus_two_cpu:/models/
half_plus_two" \
    -e MODEL_NAME=half_plus_two \
    tensorflow/serving &
```

We will see that the server has been started. We should get the following log in the terminal:

```
→    TF_SERVE 2022-09-30 13:55:35.891150: I tensorflow_
serving/model_servers/server.cc:89] Building single
*** Truncated lines ***
2022-09-30 13:55:37.710744: I tensorflow_serving/core/
basic_manager.cc:740] Successfully reserved resources to
load servable {name: half_plus_two version: 123}
2022-09-30 13:55:37.711029: I tensorflow_serving/core/
loader_harness.cc:66] Approving load for servable version
{name: half_plus_two version: 123}
*** Truncated lines ***
```

```
Reading SavedModel from: /models/half_plus_two/00000123
2022-09-30 13:55:37.727111: I external/org_tensorflow/
tensorflow/cc/saved_model/reader.cc:89] Reading meta
graph with tags { serve }
*** Truncated lines ***
2022-09-30 13:55:38.115162: I tensorflow_serving/
model_servers/server.cc:442] Exporting HTTP/REST API
at:localhost:8501 ...
```

The output is truncated to avoid some messages, but I suggest that you read all the messages to understand the process better. I have highlighted some lines in the block to show you that Serving loaded version 123 from the 0000123 folder that we saw in *Figure 12.3*. This is also obvious from the Reading SavedModel from: /models/half_plus_two/00000123 line in the preceding terminal logs. This /models/half_plus_two path is created by the **Loader** to serve the model to the client as instructed by the Manager. We passed this path when starting the server for the model as -v "$TESTDATA/saved_model_half_plus_two_cpu:/models/half_plus_two" in *step 7*.

8. Now, we will do something fun. We will just copy the 0000123 folder and give the copied folder the name 0000124. We will see both folders now, as shown in *Figure 12.5*:

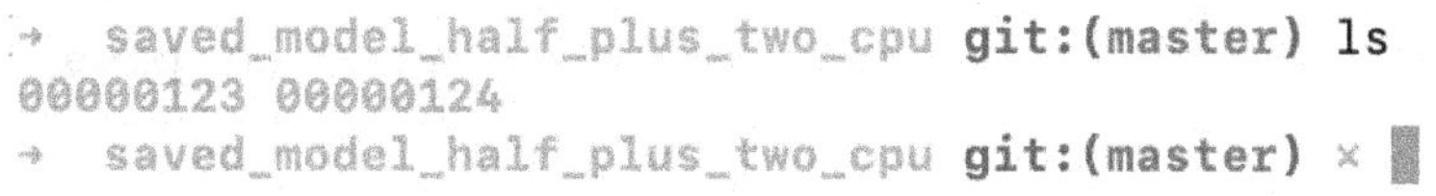

```
→  saved_model_half_plus_two_cpu git:(master) ls
00000123 00000124
→  saved_model_half_plus_two_cpu git:(master) × █
```

Figure 12.5 – A new version of the model can be created by copying the old version

9. Now, let's go to the terminal and look at the logs:

```
→  TF_SERVE 2022-09-30 15:58:32.067867: I tensorflow_
serving/core/basic_manager.cc:740] Successfully reserved
resources to load servable {name: half_plus_two version:
124}
2022-09-30 15:58:32.069668: I tensorflow_serving/core/
loader_harness.cc:66] Approving load for servable version
{name: half_plus_two version: 124}
2022-09-30 15:58:32.069743: I tensorflow_serving/core/
loader_harness.cc:74] Loading servable version {name:
half_plus_two version: 124}
2022-09-30 15:58:32.077560: I external/org_tensorflow/
tensorflow/cc/saved_model/reader.cc:45] Reading
SavedModel from: /models/half_plus_two/00000124
*** Truncated lines ***
```

```
Successfully loaded servable version {name: half_plus_two
version: 124}
*** Truncated lines ***
2022-09-30 15:58:32.379517: I tensorflow_serving/core/
loader_harness.cc:119] Unloading just-loaded servable
version {name: half_plus_two version: 123}
*** Truncated lines ***
  tensorflow_serving/core/loader_harness.cc:127] Done
unloading servable version {name: half_plus_two version:
123}
```

Again, we have truncated the logs for better reading. However, feel free to go through the full logs to view the full story. I have highlighted some lines to show you how, immediately, once the new model version is available, the Manager loads the new model first and then unloads the old model. The Manager is using the availability versioning policy here.

10. Now, let's call the inference API. Run the following command in your terminal to call the predict API:

```
curl -d '{"instances": [1.0, 2.0, 5.0]}' \
    -X POST http://localhost:8501/v1/models/half_plus_
two:predict
```

We will get the following response:

```
{
    "predictions": [2.5, 3.0, 4.5
    ]
}
```

We can also test the API call from Postman. We can send a request, as shown in *Figure 12.6*, and get the response through a more user-friendly interface:

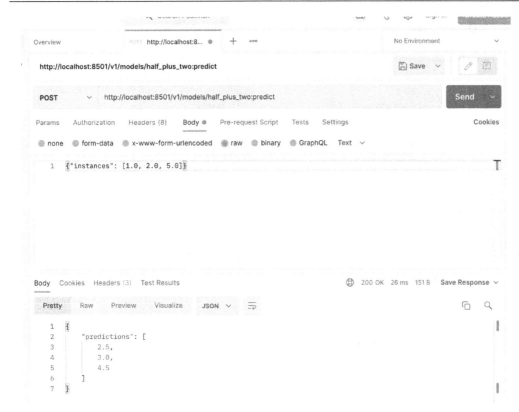

Figure 12.6 – Using Postman to send requests and get responses
from the endpoint exposed by TensorFlow Serving

If you are interested in knowing what the model that we have demonstrated can do, you can look at the source code in the file for the model at `https://github.com/tensorflow/serving/blob/master/tensorflow_serving/servables/tensorflow/testdata/saved_model_half_plus_two.py`. The code is very big because it is used to generate models for multiple platforms.

However, at a high level, it computes the $y = a*x + b$ function. Here, $a = 0.5$ and $b = 2.0$.

Now, we can look at how the output for the inputs, [1.0, 2.0, 5.0], are computed:

```
y = a*1.0 + b = 0.5 * 1.0 + 2.0 = 2.5
y = a*2.0 + b = 0.5 * 2.0 + 2.0 = 3.0
y = a*5.0 + b = 0.5 * 5.0 + 2.0 = 4.5
```

So, the final response list is [2.5, 3.0, 4.5], which combines the responses for all three input instances.

This is the response that we got from the API call, as shown in *Figure 12.6*.

Using advanced model configurations

You can also use advanced model configurations during serving. In that case, we need to provide the path to the config file. To run the server using Docker, we need to use the following command:

```
docker run -t --rm -p 8501:8501 -v "$(pwd):/
models/"  tensorflow/serving \
    --model_config_file="/models/models.config" \
    --model_config_file_poll_wait_seconds=60
```

The `models.config` file contains the config that will be used when serving. We are also setting a 60-second interval to poll the updated config from the `config` file. If anything changes in the config file, then the servables might be updated.

The following is a sample `config` file:

```
model_config_list {
  config {
    name: 'my_first_model'
    base_path: '/tmp/my_first_model/'
    model_platform: 'tensorflow'
  }
  config {
    name: 'my_second_model'
    base_path: '/tmp/my_second_model/'
    model_platform: 'tensorflow'
  }
}
```

It can contain a list of configs. From the preceding `config`, we can see that we can serve more than one model in different paths using TensorFlow Serving.

We can also specify the version of a model that will be loaded. By default, the newest version is loaded. However, we can specify the version that needs to be loaded using the following code snippet in the code:

```
Config {
 name: ' my_first_model '
 base_path: '/tmp/my_first_model/'
 model_platform: 'tensorflow'
 model_version_policy {
   specific{
```

```
        versions:123
  Versions: 124
      }
    }
  }
```

In the preceding code snippet, we specified two versions to load for the my_first_model model.

Then, in the API call, we can specify the version number with /v1/models/<model name>/versions/<version number>.

To demonstrate the use of the config file, let's run the following example. Let's go to the testdata directory using the cd serving/tensorflow_serving/servables/tensorflow/testdata command, as shown in *Figure 12.2*. Here, let's find the two servables named saved_model_half_plus_two_cpu and saved_model_half_plus_three. We will serve these models together:

1. First of all, let's create a config file called model.config:

```
model_config_list: {
  config: {
    name: "saved_model_half_plus_two_cpu",
    base_path:
    "/models/saved_model_half_plus_two_cpu/",
    model_platform: "tensorflow",
    model_version_policy {
        specific{
                versions:123
                versions: 124
            }
        }
  },
  config: {
    name: "saved_model_half_plus_three",
    base_path: "/models/saved_model_half_plus_three/",
    model_platform: "tensorflow"
  }
}
```

We have created two configs for the two models and have added support for loading two versions for the `saved_model_half_plus_two_cpu` model.

2. Now, let's start the server using the following command:

```
docker run -t --rm -p 8501:8501 -v "$(pwd):/
models/"  tensorflow/serving \
    --model_config_file="/models/model.config" \
    --model_config_file_poll_wait_seconds=60
```

Now, consider why we have used `/models` in the paths in the `config` file. This is because the current repository is located at the `/models/` location due to the `-v "$(pwd):/models/"` part of the code snippet. Knowing this will help you write `config` files without confusion in your work.

3. Now, let's look at the logs in the terminal. We have the following log:

```
2022-10-01 01:53:52.942788: I tensorflow_serving/model_
servers/server_core.cc:465] Adding/updating models.
*** Truncated lines ***
2022-10-01 01:53:53.178661: I external/org_tensorflow/
tensorflow/cc/saved_model/reader.cc:45] Reading
SavedModel from: /models/saved_model_half_plus_
three/00000123
*** Truncated lines ***
2022-10-01 01:53:53.756802: I tensorflow_serving/core/
loader_harness.cc:95] Successfully loaded servable
version {name: saved_model_half_plus_two_cpu version:
124}
2022-10-01 01:53:53.893314: I tensorflow_serving/core/
loader_harness.cc:95] Successfully loaded servable
version {name: saved_model_half_plus_two_cpu version:
123}
*** Truncated lines ***
2022-10-01 01:57:53.647799: I tensorflow_serving/model_
servers/server_core.cc:486] Finished adding/updating
models
```

From the highlighted part of the logs, we notice that one model is loaded for `saved_model_half_plus_three`, but for the `saved_model_half_plus_two_cpu` model, two different versions are loaded, as specified in the model `config` file.

4. Now, let's go to Postman and try to call the `predict` method on the models. First of all, let's call the `predict` method on the `saved_model_half_plus_three` model using Postman. The response is shown in *Figure 12.7*:

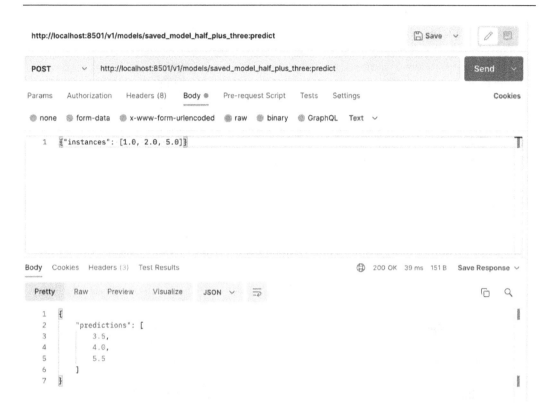

Figure 12.7 – Calling the predict method on the saved_model_half_plus_three model

We get a response of $[3.5, 4.0, 5.5]$ for the input, $[1.0, 2.0, 5.0]$. The model is simply calculating the `y = a*x + c` function, where `a = 0.5` and `c = 3`.

The responses of $[3.5, 4.0, 5.5]$ are computed in the following way:

```
y = a*x + c = 0.5*1.0 + 3 = 3.5
y = a*x + c = 0.5*2.0 + 3 = 4.0
y = a*x + c = 0.5*3.0 + 3 = 5.5
```

We did not have to specify the version number in the URL. The URL is simply `http://localhost:8501/v1/models/saved_model_half_plus_three:predict`.

However, we could also specify the version number in the URL as `http://localhost:8501/v1/models/saved_model_half_plus_three/versions/123:predict` and still get the same response.

5. Now, the other model supports multiple versions, so the user can access different models by specifying the version number. If the version number is not specified, then, by default, the latest version of the model will be accessed. For example, the URL `http://localhost:8501/v1/models/saved_model_half_plus_two_cpu:predict` will invoke the latest

version, meaning version `124` of the model. We can specify the versions in the URL as well. For example, to call version `123` of the model, we can use `http://localhost:8501/v1/models/saved_model_half_plus_two_cpu/versions/123:predict`, and to call version `124` of the model, we just skip adding the version information in the URL as `124` is the latest version.

> **Advanced configuration**
>
> To learn more about advanced configuration, please check out `https://www.tensorflow.org/tfx/serving/serving_config#model_server_configuration`.

In this section, we used TensorFlow Serving to serve a dummy model and saw how the model can be invoked using the REST API. We also saw how the model versioning update works in TensorFlow Serving. In the next section, we will draw up a summary of this chapter and then conclude.

Summary

In this chapter, we explored a popular tool for serving models. TensorFlow Serving can be used to serve any TensorFlow model, and the serving architecture handles versioning issues with high performance. We also introduced you to the architecture of TensorFlow Serving at a high level. Then, we served a model provided by the TensorFlow repository through Docker and explained how model serving is done step by step. Finally, we demonstrated how can we call the API to get the predictions from Postman.

In the next chapter, we will talk about another popular model-serving tool known as Ray Serve.

Further reading

You can read the following articles to find out more about TensorFlow Serving:

* To learn more about the TensorFlow architecture: `https://www.tensorflow.org/tfx/serving/architecture`

* To learn more about advanced model-serving configurations: `https://www.tensorflow.org/tfx/serving/serving_config`

* To learn how to use TensorFlow Serving without Docker: `https://www.tensorflow.org/tfx/serving/api_rest`

* To explore TensorFlow Serving with Kubernetes: `https://www.tensorflow.org/tfx/serving/serving_kubernetes`

13
Using Ray Serve

In this chapter, we will talk about serving models with Ray Serve. This is one of the most popular tools for serving ML models. It is a framework-agnostic scalable model-serving library. ML models created using almost any library can be served using Ray Serve. We will explore this library in this chapter and show some hands-on examples to get you up and running with Ray Serve. Covering all the topics and concepts of Ray Serve itself would demand a separate book. So, we will just cover some basic information and the end-to-end process of using Ray Serve.

At a high level, we are going to cover the following main topics in this chapter:

- Introducing Ray Serve
- Using Ray Serve to serve a model

Technical requirements

In this chapter, we will mostly use the same libraries that we used in the previous chapters. You should have Postman or another REST API client installed to be able to send API calls and see the response. All the code for this chapter is provided at `https://github.com/PacktPublishing/ Machine-Learning-Model-Serving-Patterns-and-Best-Practices/tree/ main/Chapter%2013`.

If you get `ModuleNotFoundError` while trying to import any library, then you should install that module using the `pip3 install <module_name>` command. In this chapter, you will need to install Ray Serve. Please use the `pip3 install "ray[serve]"` command to do so.

Introducing Ray Serve

Ray Serve is a framework-agnostic model-serving library. It is scalable and creates inference APIs on your behalf. Some of the key concepts in Ray Serve are as follows:

- Deployment

- ServeHandle
- Ingress deployment

We will look at each of these in the following sections.

Deployment

A deployment contains the business logic and the ML model that will be served. To define a deployment, the @serve.deployment decorator is used. For example, let's take a look at the following code snippet, which shows a very basic deployment that will return whatever message is passed by the user as a payload:

```
@serve.deployment
class MyFirstDeployment:
  # Take the message to return as an argument to the
constructor.
  def __init__(self, msg):
      self.msg = msg

  def __call__(self):
      return self.msg

my_first_deployment = MyFirstDeployment.bind("Hello world!")
```

In this code snippet, we define the MyFirstDeployment deployment using the @serve.deployment decorator. This class has the __call__ method, which is invoked whenever we hit the endpoint from the server. We can see that this method returns whatever message is passed while creating the instance of the deployment. For example, we can create a deployment instance using the following code snippet:

```
my_first_deployment = MyFirstDeployment.bind("Hello world!")
```

Now, we can serve the deployment using the following command:

```
serve run first_ray_serve_demo:my_first_deployment
```

Here, first_ray_serve_demo is the name of the Python file with the .py extension.

After running this command, we will see the following log:

```
2022-10-16 08:31:24,992 INFO scripts.py:294 -- Deploying from
import path: "first_ray_serve_demo:my_first_deployment".
```

```
2022-10-16 08:31:30,687 INFO worker.py:1509 -- Started a local
Ray instance. View the dashboard at 127.0.0.1:8265

(ServeController pid=58356) INFO 2022-10-16 08:31:34,636
controller 58356 http_state.py:129 - Starting HTTP proxy
with name 'SERVE_CONTROLLER_ACTOR:SERVE_PROXY_ACTOR-9080e55
fa7ae17885d254728a191fb44766ea3105cd0c63fb624e6bd' on node
'9080e55fa7ae17885d254728a191fb44766ea3105cd0c63fb624e6bd'
listening on '127.0.0.1:8000'

(ServeController pid=58356) INFO 2022-10-16 08:31:35,674
controller 58356 deployment_state.py:1232 - Adding 1 replicas
to deployment 'MyFirstDeployment'.

(HTTPProxyActor pid=58369) INFO:      Started server process
[58369]
```

Let's look at the highlighted portions of the code. From the first highlighted part of the log, we notice that a dashboard is created where we can see the local instance that is running. If we go to **127.0.0.1:8265**, we will see a dashboard showing all the active nodes, as shown in *Figure 13.1*:

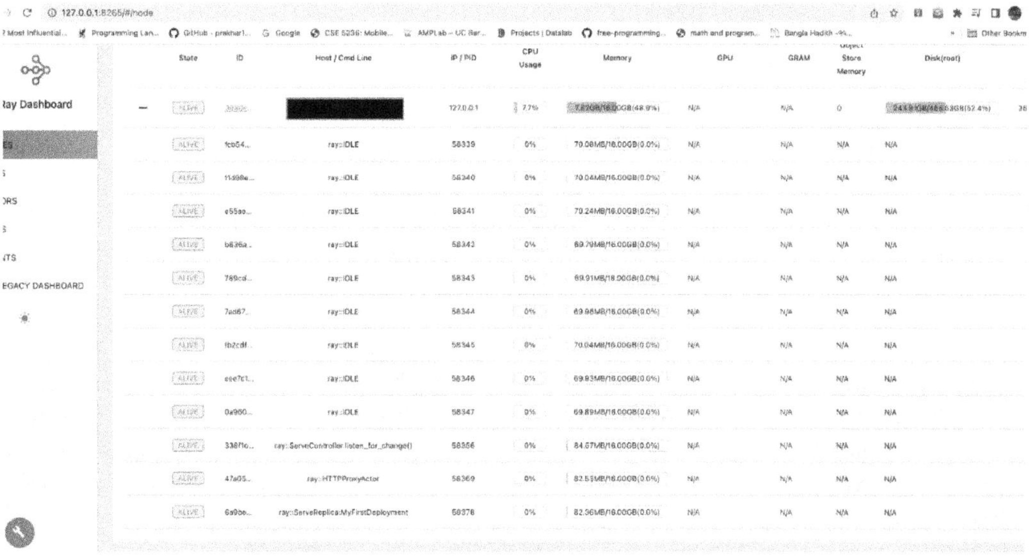

Figure 13.1 – Ray dashboard showing all the active nodes

From this figure, we notice that an active node with an ID starting with **908** is running the HTTP proxy. We also notice that a replica for our deployment, ray::ServerReplica:MyFirstDeployment, is running. From the log, we notice that the server is listening to the HTTP requests on 127.0.0.1:8000. Let's go to that endpoint from *Postman*. We will see the following response in *Postman*:

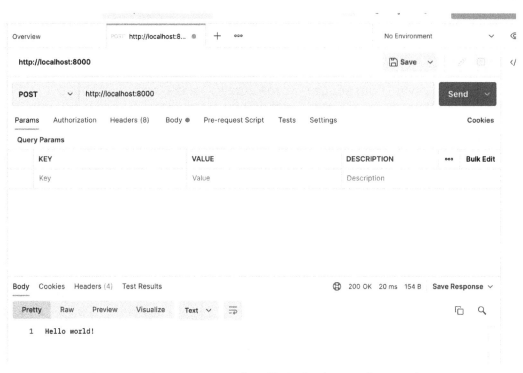

Figure 13.2 – Postman response from MyFirstDeployment ServerReplica

We got the response `Hello World!` because this is what we passed when creating the deployment instance in our code. We did not provide an option to pass the payload from the HTTP request.

ServeHandle

`ServeHandle` is a reference to a running deployment. If we call the remote method on `ServeHandle`, then we get the same response that we would get using an HTTP request. This allows us to smartly invoke the inference APIs from the program. Using `ServeHandle`, we can do a composite of multiple models. We can pass the models as ServeHandles during binding. For example, let's consider the following code snippet:

```
import ray
from ray import serve

@serve.deployment
class ModelA:
    def __init__(self):
        self.model = lambda x : x + 5

    async def __call__(self, request):
```

```
            data = await request.json()
            x = data['X']
            return self.model(x)

@serve.deployment
class ModelB:
    def __init__(self):
        self.model = lambda x : x * 2

    async def __call__(self, request):
        data = await request.json()
        x = data['X']
        return self.model(x)

@serve.deployment
class Driver:
    def __init__(self, model_a_handle, model_b_handle):
        self._model_a_handle = model_a_handle
        self._model_b_handle = model_b_handle

    async def __call__(self, request):
        ref_a = await self._model_a_handle.remote(request)
        ref_b = await self._model_b_handle.remote(request)
        return (await ref_a) + (await ref_b)


model_a = ModelA.bind()
model_b = ModelB.bind()

driver = Driver.bind(model_a, model_b)
```

In this code snippet, we have two models, ModelA and ModelB. These models just contain dummy computations for now. ModelA takes x as input and returns x + 5. ModelB takes x as input and returns 2*x. Now, note that we are using another deployment called Driver, where we are passing both ModelA and ModelB after binding. These models are passed as ServeHandles and can independently provide an HTTP response or a response using the remote() function. Now, let's run the program using the following command in the terminal:

```
serve run model_composition:driver
```

Now, we can take a look at the log of the model deployment to understand what is happening. I have taken the following snippet of the log to show that Ray Serve has created a separate deployment for each of the deployments:

```
(ServeController pid=94883) INFO 2022-10-16 15:40:12,999
```

```
controller 94883 deployment_state.py:1232 - Adding 1 replicas
to deployment 'ModelA'.
(ServeController pid=94883) INFO 2022-10-16 15:40:13,019
controller 94883 deployment_state.py:1232 - Adding 1 replicas
to deployment 'ModelB'.
(ServeController pid=94883) INFO 2022-10-16 15:40:13,034
controller 94883 deployment_state.py:1232 - Adding 1 replicas
to deployment 'Driver'.
```

Here, we have passed the ServeHandles for ModelA and ModelB to the driver. Inside the driver, we can make HTTP calls to the ServeHandles using the following snippet:

```
ref_a = await self._model_a_handle.remote(request)
ref_b = await self._model_b_handle.remote(request)
```

Now, we can make a request from Postman and see the output, as shown in *Figure 13.3*:

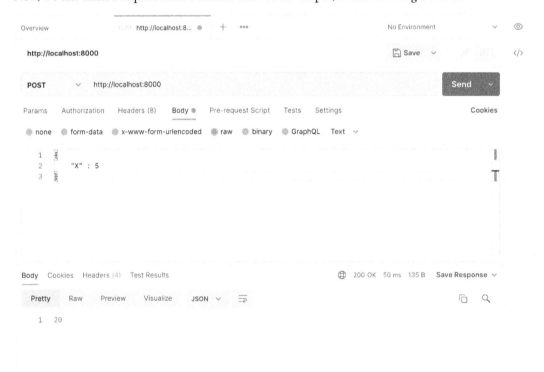

Figure 13.3 – Example of making HTTP calls to a deployment with model composition

We got an output of 20 by passing the input, { "X" : 5 }. Let's see how we get this output:

- ModelA takes the input and returns 5 + 5 = 10

- ModelB takes the input and returns 2*5 = 10

- Driver takes the responses from both models and returns the addition of these results using return (await ref_a) + (await ref_b)

Therefore, we got 20. This idea of ServeHandles is to enable you to composite multiple models and effectively carry out model serving using the ensemble pattern.

Ingress deployment

When serving multiple models using model composition, there will always be a top-level deployment that will handle HTTP requests. For example, in the preceding code snippet, Driver takes care of handling the HTTP requests. It then calls other models in the composition using ServeHandles. This top-level deployment is called ingress deployment. Remember that everything with the @serve. deployment decorator is a deployment. Other deployments can be specific to particular models and they are passed to the ingress deployment as ServeHandles after binding using the bind() method.

Deployment graph

Ray Serve supports an API to form a graph of different deployments. Using this concept, we can compose multiple models in the form of a graph. This helps us serve an ML model using the pipeline pattern. For example, let's look at the following code snippet, where we are creating a pipeline with four sequential steps. Each step of the pipeline just adds a suffix with the step name. This will help us understand in what order the steps in the graph are executed:

1. Use the following code snippet to create step 1 of the pipeline in the Step1 function:

    ```
    import ray
    from ray import serve
    from ray.serve.dag import InputNode
    from ray.serve.drivers import DAGDriver
    @serve.deployment
    def Step1(inp: str) -> str:
        print("I am inside step 1")
        return f"{inp}_step1"
    ```

    The Step1 method will create the first step in the pipeline and will print a message stating I am inside step 1. It will also return a string ending with the _step1 suffix, indicating the output is coming from step 1 of the pipeline.

2. In this code snippet, we just added the _step1 suffix to the input. The program for step 2 of the pipeline is as follows:

    ```
    @serve.deployment
    ```

```
def Step2(inp: str) -> str:
    print("I am inside step 2")
    return f"{inp}_step2"
```

The Step2 method will create step 2 of the pipeline. This step will print a message stating I am inside step 2 and return a string ending with the _step2 suffix.

3. This step adds the _step2 suffix to the input. The program for step 3 is as follows:

```
@serve.deployment
def Step3(inp: str) -> str:
    print("I am inside step 3")
    return f"{inp}_step3"
```

Inside the Step3 function, we put the logic for step 3 of the pipeline. This step prints a message stating I am inside step 3 and returns a string ending with the _step3 suffix.

4. The program for step 4 is as follows:

```
@serve.deployment
def Step4(inp: str) -> str:
    print("I am inside step 4")
    return f"{inp}_step4"
```

Step 4 adds the _step4 suffix to the input.

5. The model deployment can be created using the following class, which provides the prediction response. The predict method adds the _predict suffix to the input:

```
@serve.deployment
class Model:
    def __init__(self):
        self.model = lambda x : f"{x}_predict"

    def predict(self, inp: str) -> str:
        print("I am inside predict method")
        return self.model(inp)


with InputNode() as input:
    model = Model.bind()
```

```
    step1 = Step1.bind(input)
    step2 = Step2.bind(step1)
    step3 = Step3.bind(step2)
    step4 = Step4.bind(step3)
    output = model.predict.bind(step4)
    serve_dag = DAGDriver.bind(output)

handle = serve.run(serve_dag)
print(ray.get(handle.predict.remote("hello")))
```

Now, if we run the program, we will get `hello_step1_step2_step3_step4_predict` as output. We can run it as many times as we want and we will get the same output. So, we got the idea that the steps in the graph or pipeline are executed in order. Following this technique, we can use the pipeline pattern to deploy the ML model.

In this section, we have provided a high-level overview of Ray Serve and discussed some key concepts, along with examples. In the next section, we will use Ray Serve to serve some models from end to end. First, we will use the composition of models to serve a model while following the ensemble pattern, and then use the concept of a deployment graph to deploy a model following the pipeline pattern.

Using Ray Serve to serve a model

In this section, we will use Ray Serve to serve two dummy models using the ensemble and pipeline patterns. We will use very basic models just to demonstrate the end-to-end process of using Ray Serve.

Using the ensemble pattern in Ray Serve

In this subsection, we will use Ray Serve to ensemble the results of two scikit-learn models:

1. First of all, let's create a deployment for deploying a `RandomForestRegressor` model using the following code snippet:

```
from ray import serve
from sklearn.ensemble import RandomForestRegressor
from sklearn.ensemble import AdaBoostRegressor
from sklearn.datasets import make_regression

@serve.deployment
class RandomForestRegressorModel:
    def __init__(self):
        X, y = make_regression(n_features=4, n_
informative=2,
                               random_state=0,
shuffle=False)
```

```
        self.model = RandomForestRegressor(max_depth=2,
random_state=0)
        self.model.fit(X, y)

    async def __call__(self, request):
        data = await request.json()
        x = data['X']
        pred = self.model.predict(x)
        print("Prediction from RandomForestRegressor is
", pred)
        return pred
```

Note that inside the constructor we are training, the model is using dummy data. Ideally, you would need to import the model from persistent storage instead of training on the fly inside the constructor. We are using this dummy dataset for training as we just want to show how ensembling using Ray Serve works. In practice, to solve a real business problem, you need data from well-defined sources. We predicted with the input data inside the __call__ method. We also added a `print` statement to see the prediction from this model in the log.

2. Now, let's create another deployment for serving the AdaBoostRegressor model with the following code snippet:

```
@serve.deployment
class AdaBoostRegressorModel:
    def __init__(self):
        X, y = make_regression(n_features=4, n_
informative=2,
                                random_state=0,
shuffle=False)
        self.model = AdaBoostRegressor(random_state=0, n_
estimators=100)
        self.model.fit(X, y)

    async def __call__(self, request):
        data = await request.json()
        x = data['X']
        pred = self.model.predict(x)
        print("Prediction from AdaBoostRegressor is ",
pred)
        return pred
```

We follow the same style of training for this model as in *step 1*.

3. Next, let's create a deployment to make a composition of the preceding two models by averaging the responses from them together, as shown in the following code snippet:

```
@serve.deployment
```

```
class EnsemblePattern:
    def __init__(self, model_a_handle, model_b_handle):
        self._model_a_handle = model_a_handle
        self._model_b_handle = model_b_handle

    async def __call__(self, request):
        ref_a = await self._model_a_handle.
remote(request)
        ref_b = await self._model_b_handle.
remote(request)
        return ((await ref_a) + (await ref_b))/2.0
```

In this code snippet, we provided an argument in the constructor to pass two models as ServeHandles. These ServeHandles can be used to make an HTTP query to get responses from their models. This EnsemblePattern deployment is also known as an ingress deployment as it works as the client-facing deployment to handle HTTP requests.

4.   Now, let's provide the ingress deployment with two models as ServeHandles, as shown in the following code snippet:

```
model_a = RandomForestRegressorModel.bind()
model_b = AdaBoostRegressorModel.bind()

ensemble = EnsemblePattern.bind(model_a, model_b)
```

We now have the ingress deployment ensemble ready to be deployed. Let's save the full code inside a file named ensemble_example.py. We have to use the name of this file while serving the deployment.

5.   Now, we can start running the deployment using the following command:

```
serve run ensemble_example:ensemble
```

Here, ensemble_example is the name of the Python module and ensemble is the name of the ingress deployment. We will see the following log, indicating that the server has started and that the application is ready to take the request:

```
2022-10-16 21:53:15,083 INFO scripts.py:294 -- Deploying
from import path: "ensemble_example:ensemble".
2022-10-16 21:53:18,589 INFO worker.py:1509 -- Started a
local Ray instance. View the dashboard at 127.0.0.1:8265
(ServeController pid=16220) INFO 2022-10-
16 21:53:22,910 controller 16220 http_state.
py:129 - Starting HTTP proxy with name 'SERVE_
CONTROLLER_ACTOR:SERVE_PROXY_ACTOR-a650b61d63c7fad
e6fba77006da65d388b65d834b162db7dc7ac45b4' on node
```

```
'a650b61d63c7fade6fba77006da65d388b65d834b162db7dc7ac45b4'
listening on '127.0.0.1:8000'
(HTTPProxyActor pid=16223) INFO:
    Started server process [16223]
(ServeController pid=16220) INFO 2022-10-16 21:53:23,944
controller 16220 deployment_state.py:1232 - Adding 1
replicas to deployment 'RandomForestRegressorModel'.
(ServeController pid=16220) INFO 2022-10-16 21:53:23,962
controller 16220 deployment_state.py:1232 - Adding 1
replicas to deployment 'AdaBoostRegressorModel'.
(ServeController pid=16220) INFO 2022-10-16 21:53:23,976
controller 16220 deployment_state.py:1232 - Adding 1
replicas to deployment 'EnsemblePattern'.
2022-10-16 21:53:26,884 SUCC scripts.py:307 -- Deployed
successfully.
```

I have highlighted a portion of the log indicating that for each of the deployments, a separate replica is created. They can now operate independently and can be composed in any combination. In this way, Ray Serve is a very efficient tool in supporting the ensemble pattern.

6. Now, we can test the inference API from Postman. Let's send a request with [[0, 0, 0, 0]] as input. We will see the following output in the Postman response panel:

Figure 13.4 – Sending a request to the ensemble pattern model from Postman

Remember that we printed some logs from each of the deployments. Let's go to the console to see these logs. We will see some new logs, such as the following, after the request is submitted from Postman:

```
(HTTPProxyActor pid=16223) INFO 2022-10-16 22:02:46,112 http_
proxy 127.0.0.1 http_proxy.py:315 - POST / 200 1227.9ms
(ServeReplica:AdaBoostRegressorModel pid=16227)
INFO 2022-10-16 22:02:46,109 AdaBoostRegressorModel
AdaBoostRegressorModel#cxaWxB replica.py:482 - HANDLE __call__
OK 7.3ms
(ServeReplica:RandomForestRegressorModel pid=16226)
INFO 2022-10-16 22:02:46,099 RandomForestRegressorModel
RandomForestRegressorModel#AdWxyq replica.py:482 - HANDLE __
call__ OK 8.5ms
(ServeReplica:AdaBoostRegressorModel pid=16227) Prediction from
AdaBoostRegressor is [4.79722349]
(ServeReplica:RandomForestRegressorModel pid=16226) Prediction
from RandomForestRegressor is [-8.32987858]
(ServeReplica:EnsemblePattern pid=16228) INFO 2022-10-16
22:02:46,110 EnsemblePattern EnsemblePattern#kHmkKI replica.
py:482 - HANDLE __call__ OK 1220.7ms
```

The highlighted lines in the preceding logs show the predictions from the models. The prediction from AdaBoostRegressor is 4.79722349 and the prediction from RandomForestRegressor is -8.32987858. From the ingress deployment, we got the average of these two numbers as the final prediction – that is, (4.79722349 -8.32987858)/2 = -1.7663275448912135.

In this subsection, we showed you how to use Ray Serve to serve models using the ensemble pattern. In the next subsection, we will show you how to use Ray Serve to serve models using the pipeline pattern.

Using Ray Serve with the pipeline pattern

We discussed the pipeline pattern in *Chapter 9*. We can use the Ray Serve tool to serve models using the pipeline pattern. To use Ray Serve with the pipeline pattern, we need to use the concept of a deployment graph in Ray Serve, as we discussed previously.

We will create a pipeline by going through the following stages:

1. Data collection
2. Data cleaning
3. Feature extraction
4. Training
5. Predict

We will just use basic dummy operations in these steps to illustrate how the pipeline pattern works in Ray Serve. In the data collection stage, we will create a pandas `DataFrame` with some dummy data; in the data cleaning stage, we will do some basic cleaning to remove rows with missing data; in the feature extraction stage, we will convert the categorical data into numerical data using a built-in `pandas` method, and in the training stage, we will train a `RandomForestRegression` model with the features. Let's get started:

1. First, let's carry out the data collection step using the following code snippet:

    ```
    @serve.deployment
    def collect_data() -> pd.DataFrame:
        df = pd.DataFrame({"F1": [1, 2, 3, 4, 5, None], "F2":
    ["C1", "C1", "C2", "C2", "C3", "C3"], "Y": [0, 0, 0, 1,
    1, 1]})
        print("DF in the data collection stage")
        print(df)
        return df
    ```

    This step creates a dummy DataFrame. In models that serve real business problems, the data may come from databases, files, streaming data sources, and so on.

2. Note that in the data in *step 1*, there are some null values. We can carry out a basic data-cleaning step to remove these null values:

    ```
    @serve.deployment
    def data_cleaning(df: pd.DataFrame) -> pd.DataFrame:
        df = df.dropna()
        print("DF after the data cleaning stage")
        print(df)
        return df
    ```

    This step returns the data after removing the rows that contain null values.

3. Now, let's move on to a basic feature engineering step. In this step, we must encode the categorical values in the F2 column using the `pandas.get_dummies` API. The code snippet for this step is as follows:

    ```
    @serve.deployment
    def feature_engineering(df: pd.DataFrame) ->
    pd.DataFrame:
        df = pd.get_dummies(df)
        print("DF after the feature engineering stage")
        print(df)
        return df
    ```

After the cleaning, we return the modified DataFrame using `return df` so that the next stage in the pipeline can receive it. The next stage in the pipeline that will do model training depends on this data.

4. The fourth step is the training step. Usually, training is not part of the model-serving process, but in the pipeline pattern, it is embedded as a step in the pipeline. The main goal of serving is achieved from the previous step. So, let's train a basic `RandomForestRegressor` model using the data passed from the previous step. The code snippet for this step is as follows:

```
@serve.deployment
def train(df: pd.DataFrame) -> RandomForestRegressor:
    X = df[["F1", "F2_C1"]]
    y = df["Y"]
    print("Inside training!")
    model = RandomForestRegressor(max_depth=2, random_
state=0)
    model.fit(X, y)
    return model
```

This stage returns the model to the prediction by using `return model` after training. The prediction stage will be able to make predictions using the model provided by this `training` stage.

5. Then, we have the predict stage. In this stage, we take the model that comes from the training stage as input and also take an input, x, that is taken from an input node of the graph. We can use the following code snippet for the prediction step:

```
@serve.deployment
def predict(model: RandomForestRegressor, x) -> float:
    print("Inside predict method")
    print("Input is", x)
    return model.predict(x)
```

After this prediction step is done, we can form a pipeline using the deployment graph concept of Ray Serve.

6. Now, let's form the deployment graph by using the stages of the pipeline. The stages work as nodes of the graph; the connection that's created between the stages using the `bind` function, as shown in the following code, works as edges. Note that each of the steps is a Ray Serve deployment with the `@serve.deployment` decorator:

```
with InputNode() as input:
    data_collection = collect_data.bind()
    clean_data = data_cleaning.bind(data_collection)
    feature_creation = feature_engineering.bind(clean_
data)
```

```
train_model = train.bind(feature_creation)
output = predict.bind(train_model, input)
serve_dag = DAGDriver.bind(output)

handle = serve.run(serve_dag)
```

The `handle` graph that was created via `handle = serve.run(serve_dag)` is now ready to run and make predictions. We can send a request from the Python program like so:

```
print(ray.get(handle.predict.remote([[1 , 1]])))
```

This line of code will make a prediction for the input of `[[1, 1]]` and print the prediction response from the model. After running the preceding program, we will see the following output in the console:

```
(ServeReplica:collect_data pid=66004) DF in the data
collection stage
(ServeReplica:collect_data pid=66004)      F1   F2  Y
(ServeReplica:collect_data pid=66004) 0  1.0  C1   0
(ServeReplica:collect_data pid=66004) 1  2.0  C1   0
(ServeReplica:collect_data pid=66004) 2  3.0  C2   0
(ServeReplica:collect_data pid=66004) 3  4.0  C2   1
(ServeReplica:collect_data pid=66004) 4  5.0  C3   1
(ServeReplica:collect_data pid=66004) 5  NaN  C4   1
```

From the preceding log, notice that it just printed the DataFrame that we created and printed. The console log from the data cleaning step is as follows:

```
(ServeReplica:data_cleaning pid=66005) DF after the data
cleaning stage
(ServeReplica:data_cleaning pid=66005)      F1   F2  Y
(ServeReplica:data_cleaning pid=66005) 0  1.0  C1   0
(ServeReplica:data_cleaning pid=66005) 1  2.0  C1   0
(ServeReplica:data_cleaning pid=66005) 2  3.0  C2   0
(ServeReplica:data_cleaning pid=66005) 3  4.0  C2   1
(ServeReplica:data_cleaning pid=66005) 4  5.0  C3   1
```

From the log in the data cleaning step, we can see that the row containing NaN is removed. The log from the feature engineering step is as follows:

```
(ServeReplica:feature_engineering pid=66006) DF after the
feature engineering stage
```

```
(ServeReplica:feature_engineering
pid=66006)       F1  Y  F2_C1  F2_C2  F2_C3
(ServeReplica:feature_engineering pid=66006)
0  1.0  0      1      0      0
(ServeReplica:feature_engineering pid=66006)
1  2.0  0      1      0      0
(ServeReplica:feature_engineering pid=66006)
2  3.0  0      0      1      0
(ServeReplica:feature_engineering pid=66006)
3  4.0  1      0      1      0
(ServeReplica:feature_engineering pid=66006)
4  5.0  1      0      0      1
```

Note that the data is encoded in the feature engineering step. We can also see features such as F2_C1, F2_C2, and F2_C3, which are created by pandas.get_dummies. The log from the training step is as follows:

```
(ServeReplica:train pid=66007) Inside training!
```

This is just a single line indicating that we have reached this step. The log from the predict method is as follows:

```
(ServeReplica:predict pid=66008) Inside predict method
(ServeReplica:predict pid=66008) Input is [[1, 1]]
```

This indicates that the input has been received correctly by the predict method.

The output from making the inference call through the print(ray.get(handle.predict.remote([[1 , 1]]))) print statement is as follows:

```
[0.02]
```

Therefore, we can see that the basic ML pipeline is working from end to end.

In this section, we have shown examples of using Ray Serve to serve ML models using the ensemble and pipeline patterns. Next, we will summarize this chapter and conclude.

Summary

In this chapter, we discussed Ray Serve. Ray Serve is a tool with a massive amount of features. Our goal was not to introduce everything about this tool in this chapter. Rather, we wanted to give you an introduction to the tool and outline the basic requirements to understand how this tool can be used in serving production-ready models while following different patterns of the serving model.

We provided an introduction to the key concepts of the tools, along with examples. We then used Ray Serve to serve two dummy models from end-to-end using the ensemble pattern and the pipeline pattern. Using these examples, we saw how Ray Serve can help you set up model serving while following the standard patterns of serving ML models.

In the next chapter, we will introduce you to `BentoML`, another tool for serving ML models.

Further reading

You can read the following articles to learn more about Ray Serve:

- To learn more about Ray Serve, check out the official documentation at `https://docs.ray.io/en/latest/serve/index.html`
- To learn more about the key concepts in Ray Serve, please use the following link: `https://docs.ray.io/en/latest/serve/key-concepts.html`

14
Using BentoML

In this chapter, we will talk about BentoML, another popular tool for deploying ML models. BentoML helps to continuously deploy ML models, and monitors and provides prediction APIs. Throughout this chapter, we will learn about the basic key concepts of BentoML and then see an end-to-end example of deploying a model using BentoML.

At a high level, we are going to cover the following main topics in this chapter:

- Introducing BentoML
- Using BentoML to serve a model

Technical requirements

In this chapter, we will mostly use the same libraries that we have used in previous chapters. You should have Postman or another REST API client installed to be able to send API calls and see the response. All the code for this chapter is provided at this link: `https://github.com/PacktPublishing/Machine-Learning-Model-Serving-Patterns-and-Best-Practices/tree/main/Chapter%2014`.

If `ModuleNotFoundError` appears while trying to import any library, then you should install the missing module using the `pip3 install <module_name>` command. You will need the `bentoml` library for this chapter. Please install `bentoml` using the `pip3 install bentoml` command.

Introducing BentoML

BentoML is a popular tool for serving ML models. It provides support for deploying models created using almost all the popular libraries. Throughout this section, we will discuss how to get started with BentoML and how to use it for serving, along with some key concepts.

We will discuss the following concepts that are needed to use BentoML:

- Preparing models

- Services and APIs

- Bento

Let's discuss each concept in detail.

Preparing models

A trained ML model cannot be directly served using BentoML because BentoML needs to convert all the models into a common format so that it can extend support to any models from any ML library. The model needs to be saved using the BentoML API. BentoML provides the save_model API for almost all the popular ML libraries. For example, if you develop an ML model using the scikit-learn library, then you need to use the bentoml.sklearn.save_model(...) API to save the model for serving using BentoML. Let's look at the following example.

1. First, we create a basic model and save it using the BentoML API:

```
import bentoml
from sklearn.ensemble import RandomForestRegressor
from sklearn.datasets import make_regression
X, y = make_regression(n_features=4, n_informative=2,
                        random_state=0, shuffle=False)
regr = RandomForestRegressor(max_depth=2, random_state=0)
regr.fit(X, y)
saved_model = bentoml.sklearn.save_model(
    name = "DummyRegressionModel",
    model = regr
)


print(saved_model)
```

After running the code, we see the following output in the console:

```
converting 'DummyRegressionModel' to lowercase:
'dummyregressionmodel'
Model(tag="dummyregressionmodel:dudusucwz6ej6ktz")
```

Let's have a look at the log, which says that the model's name has been converted to lowercase. The output from the print statement also printed the model object reference in the model store. Note that the tag of the model starts with the model name that we passed.

2. Now, let's run the `bentoml models list` command in the terminal or console, and we will see the following output:

```
(venv) (base)  % bentoml models list
Tag                                  Module
           Size          Creation Time
dummyregressionmodel:dudusucwz6ej6ktz  bentoml.sklearn
74.99 KiB  2022-10-28 09:45:06
```

So, we got a list of the models that are currently present in the list of the model store.

3. Let's run the program again, and the same model will be saved with a different tag. If we run the `bentoml models list` command in the terminal again, we will see the following output as shown in *Figure 14.1*:

Figure 14.1 – The latest model is placed earlier in the list

Let's take a look at the output shown inside the red square. This time, we have another model, and the latest model is shown at the top.

4. Let's run the program again by changing `regr = RandomForestRegressor(max_depth=2, random_state=0)` to the `regr = RandomForestRegressor(max_depth=3, random_state=0)` line. We have changed `max_depth` from 2 to 3 so that we can notice clearly which model we are loading during serving.

5. Now, from the program, let's try to load the model from the model store using the following code snippet:

```
import bentoml
from sklearn.ensemble import RandomForestRegressor
regr: RandomForestRegressor = bentoml.sklearn.load_
model("dummyregressionmodel")
print(regr)
```

We get the following output in our console from the `print` statement:

`RandomForestRegressor(max_depth=3, random_state=0)`

That means `load_model`, by default, loaded the latest model, as the `max_depth` value of the model is 3, which was changed during the last run. From there, we can understand how `BentoML` can help us load the latest version of the model deployed to the model store.

6. Now, if you want to load a specific version of the model, you have to provide the full tag name. For example, let's try to load the model with the `dummyregressionmodel:dudusucwz6ej6ktz` tag using the following code snippet:

```
import bentoml
from sklearn.ensemble import RandomForestRegressor
regr: RandomForestRegressor = bentoml.sklearn.load_
model("dummyregressionmodel:dudusucwz6ej6ktz")
print(regr)
```

The output we get this time is the following:

`RandomForestRegressor(max_depth=2, random_state=0)`

Therefore, we have loaded the earlier model where `max_depth` was 2. In this way, we can roll back to an old model if needed during serving by specifying the older version as in this code snippet: `bentoml.sklearn.load_model("dummyregressionmodel:dudusucwz6ej6ktz")`. BentoML will automatically make the rollback without any additional developer effort. Rollback is very important in a production environment. If something goes wrong, we might need to roll back to a previous stable version of the model.

Services and APIs

`BentoML` is based on the idea of service-oriented architecture, and the core building block of `BentoML` in which a user defines the model serving logic is called a *service*. All clients, such as business analysts, data scientists, and so on, who want to get inference from the served model will call a service through the APIs exposed by it. A service is created by calling `bentoml.Service()`. Let's run an example:

1. Let's create a service to serve the dummy regression model we created before, using the following code snippet:

```
import bentoml
import numpy as np
from bentoml.io import NumpyNdarray
regr_runner = bentoml.sklearn.get("dummyregressionmodel").
to_runner()
print(regr_runner)
```

```
service = bentoml.Service("DummyRegressionService",
runners=[regr_runner])

@service.api(input=NumpyNdarray(), output=NumpyNdarray())
def predict(input: np.ndarray) -> np.ndarray:
    print("input is ", input)
    response = regr_runner.run(input)
    print("Response is ", response)
    return response
```

We have created a service and given it a name, DummyRegressionService.

We have also loaded the model as a runner using the regr_runner = bentoml. sklearn.get("dummyregressionmodel").to_runner() line and used this runner during the creation of the service. Then, we created a prediction API and specified the input and output type using the @service.api(input=NumpyNdarray(), output=NumpyNdarray()) decorator.

2. Now, we can run the service using the following command in the terminal: (venv) (base) johirulislam@Johiruls-MBP BentoMLExamples % bentoml serve service1.py.

Let's look at the log to see what is happening. It is very helpful to study the log to understand what the server is doing. The log that we get in the terminal is the following:

```
Runner(runnable_class=<class 'bentoml._internal.
frameworks.sklearn.get_runnable.<locals>.
SklearnRunnable'>, runnable_init_
params={}, name='dummyregressionmodel',
models=[Model(tag="dummyregressionmodel:6ish6osw2c7uuktz",
path="/Users/johirulislam/bentoml/models/
dummyregressionmodel/6ish6osw2c7uuktz")], resource_
config=None, runner_methods=[RunnerMethod(runner=...,
name='predict', config=RunnableMethodConfig(batchable=
False, batch_dim=(0, 0), input_spec=None, output_
spec=None), max_batch_size=100, max_latency_ms=10000)]
2022-10-28T11:04:18-0500 [WARNING] [cli]
converting DummyRegressionService to lowercase:
dummyregressionservice
*** Truncated ***
```

Let's look at the highlighted lines. On the first highlighted line, note that the latest version of the model was loaded. We can verify this by running the bentoml command, bentoml models list, in the terminal to get the models list, as follows:

```
Tag                                          Module
Size           Creation Time
dummyregressionmodel:6ish6osw2c7uuktz   bentoml.sklearn
124.86 KiB   2022-10-28 09:58:13
dummyregressionmodel:3fqunzswz6b4qktz   bentoml.sklearn
74.99 KiB    2022-10-28 09:50:22
dummyregressionmodel:dudusucwz6ej6ktz   bentoml.sklearn
74.99 KiB    2022-10-28 09:45:06
```

Note that our latest model has the dummyregressionmodel:6ish6osw2c7uuktz tag. Let's also have a look at the log to see how the model is stored in the model store. The model is stored in the ~/bentoml/models/dummyregressionmodel/6ish6osw2c7uuktz path. We can see that the tag after the model name is created as a folder, under the folder created with the model's name. Let's open the ~/bentoml/models/dummyregressionmodel path, and we will see a folder structure like the one shown in *Figure 14.2*.

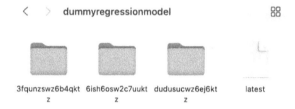

Figure 14.2 – The directory structure of the model store for the dummyregressionmodel model

3. Note that for each of the tags, there is a folder from which the model is accessed by the service. We can see a special file called latest. Let's open the file, and we will see that it contains the tag of the latest model, as shown in *Figure 14.3*.

Figure 14.3 – The latest file contains the tag of the latest model

This is the file that BentoML uses internally to keep track of the latest model. If we look at the serving log, we will see that the model with the tag shown in *Figure 14.2* was loaded earlier and created a runner for serving.

We have seen that a service takes runners that serve requests, so a service can take multiple models as runners to make predictions. This will help us to combine multiple models. A runner is a unit of serving logic.

We have seen, by default, that the name of the function with the @service.api(..) decorator becomes the API route. The name of the function in the service shown previously in the code snippet in *step 1* is predict. Therefore, predict is the endpoint for the API. We will use the Postman REST API client to send the request and get the response. If you do not have Postman installed, please read the *Technical requirements* section for instructions on how to install it. Now, let's go to Postman and try to send a request to the service API using the /predict route. The request and response from Postman are shown in *Figure 14.4*.

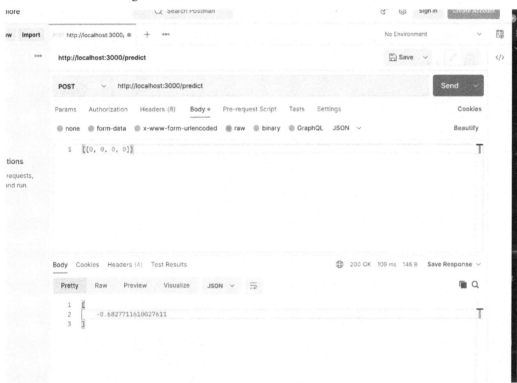

Figure 14.4 – Sending a request to dummyregressionservice using Postman and getting a response

However, if we want, we can also change the route according to our choice. For example, let's specify the route using the following code snippet:

```
@service.api(
    input=NumpyNdarray(),
    output=NumpyNdarray(),
    route="/infer"
)
def predict(input: np.ndarray) -> np.ndarray:
```

```
print("input is ", input)
response = regr_runner.run(input)
print("Response is ", response)
return response
```

We have specified the endpoint for the `predict` method as `/infer`. Now, if we go to Postman, we can send a request to the new endpoint, as shown in *Figure 14.5*.

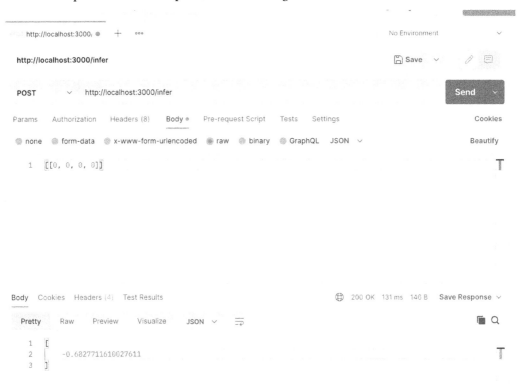

Figure 14.5 – We can specify custom routes and use them when making API calls from Postman

We can also add basic validators when we create the route. For example, let's specify the input shape during the creation of the API using the following code snippet:

```
@service.api(
    input=NumpyNdarray(
        shape=(1, 4),
        enforce_shape=True
```

```
    ),
    output=NumpyNdarray(),
    route="/infer"
)
def predict(input: np.ndarray) -> np.ndarray:
    print("input is ", input)
    response = regr_runner.run(input)
    print("Response is ", response)
    return response
```

We have specified the input shape to be (1, 4) and set enforce_shape=True. This means if the shape constraint is violated, we will get an exception. Now, let's try to send a request to the API using a numpy array of a (2, 4) shape from Postman. We get a response, as shown in *Figure 14.6*.

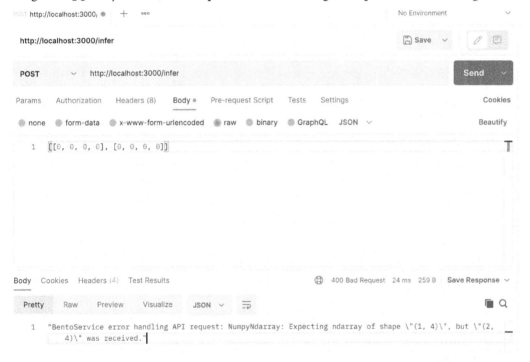

Figure 14.6 – A demonstration of the violation of validation constraints in BentoML in Postman

The error message clearly specifies that the API expects an array of a (1, 4) shape, but it got an array of a (2, 4) shape. Please note that the input of the (2, 4) shape is valid for the RandomForestRegression model we have; in fact, it can take any array of a (, 4) shape. The first number in the tuple can be any number. We just enforced this constraint for the demonstration of this validation from BentoML. This way, BentoML can provide you with some support for a business logic pattern. To learn more about different kinds of business logic in business logic patterns, please refer to *Chapter 11*, *Business Logic Pattern*.

Bento

Now, we will talk about the most important concept of BentoML, called a Bento. A Bento is an archive of files that contains the source code, models, data files, and all the dependencies needed for running a service. It standardizes how to reproduce the environment required to serve bentoml. Service.

1. To create a Bento, first, you need a special file called bentofile.yaml. It has some similarities to the build files that software engineers use. Let's create bentofile.yaml using the following contents:

   ```
   service: "bento_service_dummy:bento_service"  # Same as
   the argument passed to `bentoml serve`
   labels:
       owner: johir
       stage: dev
   include:
   - "*.py"  # A pattern for matching which files to include
   in the bento
   exclude:
   - "*.py" # A pattern for matching which files to exclude
   in the bento
   python:
       packages:  # Additional pip packages required by the
   service
       - scikit-learn
       - numpy
   ```

   Here, the "bento_service_dummy:bento_service" service contains the name of the Python file containing the service and the name of the service. We have created the file and the service named bento_service_dummy.py and bento_service respectively, as shown in *Figure 14.7*.

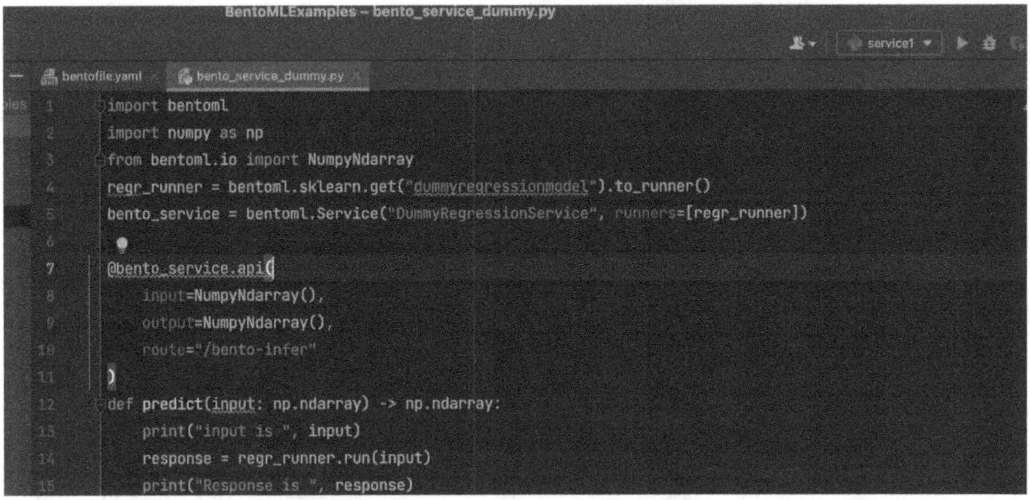

```
BentoMLExamples – bento_service_dummy.py

                                                         service1

  bentofile.yaml       bento_service_dummy.py

1    import bentoml
2    import numpy as np
3    from bentoml.io import NumpyNdarray
4    regr_runner = bentoml.sklearn.get("dummyregressionmodel").to_runner()
5    bento_service = bentoml.Service("DummyRegressionService", runners=[regr_runner])
6
7    @bento_service.api(
8        input=NumpyNdarray(),
9        output=NumpyNdarray(),
10       route="/bento-infer"
11   )
12   def predict(input: np.ndarray) -> np.ndarray:
13       print("input is ", input)
14       response = regr_runner.run(input)
15       print("Response is ", response)
```

Figure 14.7 – The Python file and the service that is used in the bentofile.yaml file

Here, the `include` field includes the files that need to be included in the Bento and the `exclude` field indicates the files that need to be excluded from the Bento.

2. Now, we can build the Bento using the `bentoml build` command in the terminal. We will see the following log in the terminal:

```
converting DummyRegressionService to lowercase:
dummyregressionservice
Building BentoML service
"dummyregressionservice:byi5uecw46p4cktz" from build
context "<root path to your programs>/BentoMLExamples"
Packing model "dummyregressionmodel:6ish6osw2c7uuktz"
Locking PyPI package versions..
  *** Truncated ***
Successfully built
Bento(tag="dummyregressionservice:byi5uecw46p4cktz")
```

Let's look at the log, and we can see that a Bento has successfully been built with the `dummyregressionservice:byi5uecw46p4cktz` tag. Now, if we go to the `/Users/<Replace with your compute name>/bentoml/` directory, we will see it has a special folder now, `bentos`, as shown in *Figure 14.8*. The username used here is the username on my computer; it will be different on your computer.

Figure 14.8 – Bentos are stored in a separate store under the bentoml directory

Let's go inside the Bento folder, dummyregressionservice, where we will find a folder with the byi5uecw46p4cktz tag name, which is shown in the log, and a file called latest, which contains the latest tag and indicates the latest version of the model.

> **Note**
>
> Model tags such as byi5uecw46p4cktz will be totally different in your case, as these are generated by the BentoML library randomly.

Now, let's go inside the byi5uecw46p4cktz directory. We will see that all the source code, required models for the service, environment information, and APIs are present inside the folder, as shown in *Figure 14.9*.

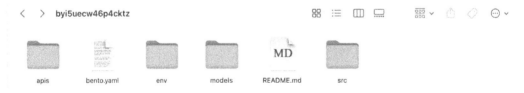

Figure 14.9 – Inside the byi5uecw46p4cktz Bento directory

Now, let's serve the Bento using the bento serve command in the terminal. Let's look at the log, where we will see the following:

```
(venv) (base) johirulislam@Johiruls-MBP BentoMLExamples %
bentoml serve
2022-10-28T12:58:16-0500 [WARNING] [cli] converting
DummyRegressionService to lowercase: dummyregressionservice
2022-10-28T12:58:16-0500 [INFO] [cli] Prometheus metrics
```

```
for HTTP BentoServer from "." can be accessed at http://
localhost:3000/metrics.
2022-10-28T12:58:16-0500 [INFO] [cli] Starting development HTTP
BentoServer from "." running on http://0.0.0.0:3000 (Press
CTRL+C to quit)
2022-10-28T12:58:17-0500 [WARNING] [dev_api_server] converting
DummyRegressionService to lowercase: dummyregressionservice
```

Note that the server is started and is ready to take requests. From Postman, we can now send a request to the service, using Bento in the same way as we did for a standalone service. For example, our endpoint for bento_service is "/bento-infer", and the request from Postman will look like *Figure 14.10*.

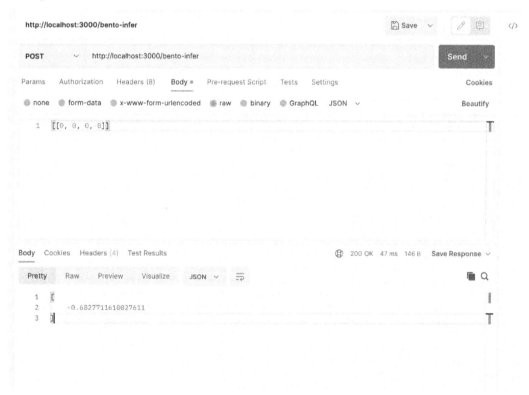

Figure 14.10 – Sending a request to Bento using Postman

The advantage of Bento over a standalone service is that Bento can be ported without the challenges of worrying about whether necessary libraries are present on the server, whether the necessary environment is set up on the server, and so on.

In this section, we have introduced you to some of the basic concepts of BentoML and have also shown some examples. In the next section, we will serve two models following the ensemble pattern of serving using BentoML.

Using BentoML to serve a model

In this section, we will serve two regression models and pass them to a service as runners. We might need to use more than one regression model to get more accurate predictions using an ensemble of multiple models. Then, we will get predictions from these two models and send the combined response to the user:

1. First of all, let's create the two models and save them using the BentoML API. First, let's create a RandomForestRegression model using the following code snippet and save it using the BentoML API:

    ```
    X, y = make_regression(n_features=4, n_informative=2,
                            random_state=0, shuffle=False)
    rf = RandomForestRegressor(max_depth=3, random_state=0)
    rf.fit(X, y)
    rf_model = bentoml.sklearn.save_model(
        name = "rf",
        model = rf
    )
    ```

    Then, we create an AdaBoostRegressor model and save it with the BentoML API using the following code snippet:

    ```
    boost = AdaBoostRegressor(random_state=0)
    boost.fit(X, y)
    boost_model = bentoml.sklearn.save_model(
        name = "boost",
        model = boost
    )
    ```

    Then, we run the program and follow the logs to ensure the models are created and converted using the BentoML API successfully. We see the following log in the terminal:

    ```
    Model(tag="rf:fmb5uxcw76xuuktz")
    Model(tag="boost:fnzdy2cw76xuuktz")
    ```

    With that, the models are saved and assigned some tags to be loaded by the BentoML service.

2. We create a service using two runners that can be used to ensemble the response from the two models. In the following code snippet, we create two runners for the two models we created before in *step 1*:

```
rf_runner = bentoml.sklearn.get("rf").to_runner()
boost_runner = bentoml.sklearn.get("boost").to_runner()
reg_service = bentoml.Service("regression_service",
runners=[rf_runner, boost_runner])

@reg_service.api(
    input=NumpyNdarray(),
    output=NumpyNdarray(),
    route="/infer"
)
def predict(input: np.ndarray) -> np.ndarray:
    print("input is ", input)
    rf_response = rf_runner.run(input)
    print("RF response ", rf_response)
    boost_response = boost_runner.run(input)
    print("Boost response ", boost_response)
    avg = (rf_response + boost_response) / 2
    print("Average is ", avg)
    return avg
```

We have saved the code along with the relevant imports in a file named `regression_service.py`.

3. We can now start the service using the `bentoml serve regression_service.py` command. We can now look at the log where we can see that the server has started and is ready to receive API calls:

```
2022-10-28T15:36:35-0500 [INFO] [cli] Prometheus metrics
for HTTP BentoServer from "regression_service.py" can be
accessed at http://localhost:3000/metrics.
2022-10-28T15:36:35-0500 [INFO] [cli] Starting
development HTTP BentoServer from "regression_service.py"
running on http://0.0.0.0:3000 (Press CTRL+C to quit)
```

Note that the server is listening on port 3000 in `localhost`.

4. Now, let's go to Postman and send an inference request from there, and we will see an output like that in *Figure 14.11*.

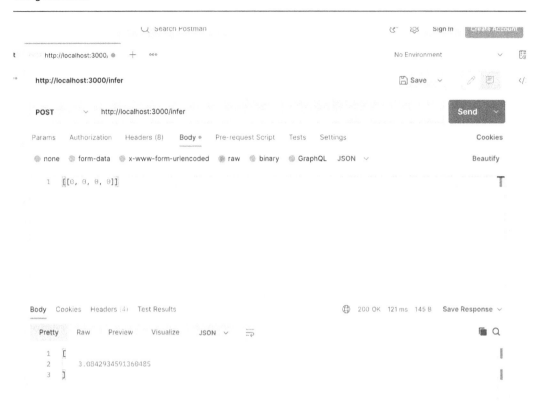

Figure 14.11 – Sending a request from Postman to the service serving
two regression models, following an ensemble pattern

Now, let's look at the log again. We should be able to see the response from the `print` statements in the log now, as shown here:

```
input is   [[0 0 0 0]]
RF response   [-0.68277116]
Boost response   [6.85135808]
Average is   [3.08429346]
2022-10-28T15:37:10-0500 [INFO] [dev_api_
server:regression_service] 127.0.0.1:62121
(scheme=http,method=POST,path=/
infer,type=application/json,length=14)
(status=200,type=application/json,length=20) 53.526ms
(trace=e0db708558d83c3bba0627f6e50b9165,
span=75abbc1f48585e7d,sampled=0)
```

From the log, we can see that the service got the correct input that we sent. The response from the `RandomForestRegression` model is `[-0.68277116]`, and the response from the `AdaBoostRegression` model is `[6.85135808]`. It took the average of these two models' responses and returned `[3.08429346]` to the caller, which is what we saw from Postman.

In this section, we have used BentoML to serve two models using an ensemble pattern. We have tried to show an end-to-end process of serving the model using a standalone service. However, during deployment to an actual production server, you should use Bento instead of a standalone service.

Summary

In this chapter, we introduced you to BentoML, a popular framework for serving machine learning models. We have shown how can you convert models into a Bento-supported format using the BentoML API. We have also shown you how can you create a service that can be served by exposing API endpoints to customers. You have also learned about Bento and have seen how it can help you to create an environment-independent service by packaging all the dependencies and data files inside a Bento.

In the next chapter, we will look at a cloud-based service that can help to serve a model.

Further reading

You can read the following articles to learn more about BentoML:

- Installation of BentoML: https://docs.bentoml.org/en/latest/installation.html

- Main concepts of BentoML: https://docs.bentoml.org/en/latest/concepts/index.html

- To learn more about the supported libraries and frameworks in BentoML, please follow this link: https://docs.bentoml.org/en/latest/frameworks/index.html

Part 4:
Exploring Cloud Solutions

In this part, we explore how we can use fully managed cloud solutions to serve machine learning models, and we demonstrate the use of Amazon Sagemaker as a representative of existing fully managed cloud solutions.

This part contains the following chapters:

## 15

Serving ML Models Using a Fully Managed Cloud Solution

We have, so far, looked at some tools such as TensorFlow Serving, Ray Serve, and BentoML. In this chapter, we will see how we can serve an **ML** model using a fully managed cloud solution. There are a few popular cloud solutions provided by Amazon, Google, Microsoft, IBM, and more. In this chapter, we will discuss how we can serve an ML model using Amazon SageMaker. Explaining all aspects of Amazon SageMaker (`https://aws.amazon.com/sagemaker/`) is beyond the scope of this book. We will only introduce some basic concepts, deploy a model, and test whether we can invoke the model.

At a high level, we are going to cover the following main topics in this chapter:

- Introducing Amazon SageMaker
- Using Amazon SageMaker to serve a model

Technical requirements

In this chapter, we will mostly use Amazon SageMaker, so you will need an AWS account for that. If you do not already have an AWS account, feel free to create one by following this link: `https://aws.amazon.com/free/`. This link will guide you on how you can create a free account for learning and trial purposes.

Introducing Amazon SageMaker

In this section, we will introduce Amazon SageMaker to demonstrate how a fully managed cloud solution can help you to serve ML models.

Amazon SageMaker is a full stack solution to ML. It helps at every step of the ML pipeline, such as feature engineering, training, tuning, deploying, and monitoring. It supports almost all the leading ML frameworks, including the following:

- TensorFlow

- PyTorch

- scikit-learn

We can create models using our chosen library and train and serve them using Amazon SageMaker. At a high level, Amazon SageMaker provides the following utilities for ML practitioners:

- Easier access to the development of ML for more people by providing IDEs and built-in no-code interfaces for business analysts

- Support to store, preprocess, and extract features from a large volume of structured and unstructured data

- An optimized framework supports faster training by reducing the training time of complex models from hours to minutes

- Automate the end-to-end MLOps process to build, train, deploy, and manage ML models at scale

Amazon SageMaker has many built-in features for supporting full stack ML solutions and utilities. We will discuss some of those features in the following subsection.

Amazon SageMaker features

In this subsection, we will talk about some of the main features of Amazon SageMaker. A detailed discussion of all these features is beyond the scope of this book. We will just give a high-level overview. Amazon SageMaker has the following built-in features as per the documentation, which can be found at `https://docs.aws.amazon.com/sagemaker/latest/dg/whatis.html`:

- **SageMaker Studio**: This is an ML IDE where you can perform the operations needed at different stages in the ML life cycle, starting from building the model and going all the way through to deploying it. You can read about SageMaker Studio in more detail by following this link: `https://docs.aws.amazon.com/sagemaker/latest/dg/studio-ui.html`.

- **SageMaker Canvas**: This provides support to build a model and get predictions without any coding. You need to import your data to the canvas, the canvas will internally make the models ready for you, and then you can make the predictions. To learn more about the tool, please follow this link: `https://docs.aws.amazon.com/sagemaker/latest/dg/canvas.html`.

- **SageMaker Ground Truth Plus**: This tool provides support to build a training dataset by labeling the data. You need to upload the data and labeling requirements and the tool will provide the labeled training data for you. This labeling task is often done manually or using some algorithms developed by data scientists. This tool makes ML easier by removing a lot of the effort. To learn more about the tool, please follow this link: `https://docs.aws.amazon.com/sagemaker/latest/dg/gtp.html`.

- **Amazon SageMaker Studio Lab**: This is a free service that provides access to Amazon computing resources from an environment based on the open source JupyterLab. This is a light version of Amazon SageMaker Studio with reduced functionalities. To learn more about this free service, please follow this link: `https://docs.aws.amazon.com/sagemaker/latest/dg/studio-lab.html`.

- **Amazon SageMaker Training Compiler**: This is used to train a deep learning model very fast on GPU instances that are managed by Amazon SageMaker. To learn more about the tool, please follow this link: `https://docs.aws.amazon.com/sagemaker/latest/dg/training-compiler.html`.

- **SageMaker serverless endpoints**: This provides serverless endpoints for hosting models and also scales automatically to serve increased traffic. To learn more about this feature, please follow this link: `https://docs.aws.amazon.com/sagemaker/latest/dg/serverless-endpoints.html`.

- **SageMaker Inference Recommender**: This tool recommends the inference instance and configuration that can be served to a live inference endpoint. This recommender helps you to find the best instance of the inference out of multiple instances present. To get a detailed understanding of this feature, please follow this link: `https://docs.aws.amazon.com/sagemaker/latest/dg/inference-recommender.html`.

- **SageMaker model registry**: This helps you to create a catalog of production models, manage the versioning of models, manage the approval stage of a model, deploy a model to production, and so on. This model registry can help you keep track of the history and evolution of a model over time, providing better audit and monitoring support. To learn more about this feature, please follow this link: `https://docs.aws.amazon.com/sagemaker/latest/dg/model-registry.html`.

- **SageMaker projects**: This helps you to automate MLOps, which means automating the end-to-end process of the ML model life cycle using CI/CD. To learn more about this feature, please follow this link: `https://docs.aws.amazon.com/sagemaker/latest/dg/sagemaker-projects.html`.

- **SageMaker Model Building Pipelines**: This helps you to create ML pipelines. To learn more about this tool, please follow this link: `https://docs.aws.amazon.com/sagemaker/latest/dg/pipelines.html`.

- **SageMaker ML Lineage Tracking**: This feature helps you to create and store information about each step in the ML workflow, starting from data preparation to deployment. With this information, we can reproduce an ML workflow, track a model, track the lineage of a dataset, and so on. To learn more about this feature, please follow this link: `https://docs.aws.amazon.com/sagemaker/latest/dg/lineage-tracking.html`.

- **SageMaker Data Wrangler**: This feature provides an end-to-end data engineering solution, such as importing, cleaning, and transforming data, extracting features, analyzing, and so on.

To learn more about this feature, please follow this link: `https://docs.aws.amazon.com/sagemaker/latest/dg/data-wrangler.html`.

- **SageMaker Feature Store**: You can store features created from the raw data using this functionality provided by SageMaker so that the features can be accessed whenever needed. This helps with the reuse of features instead of creating them from scratch all the time. To learn more about this feature, please follow this link: `https://docs.aws.amazon.com/sagemaker/latest/dg/feature-store.html`.

- **SageMaker Clarify**: This is a great feature for explainable AI within SageMaker. This will help you to remove potential bias and explain the predictions from models. Please go to `https://docs.aws.amazon.com/sagemaker/latest/dg/clarify-fairness-and-explainability.html` to learn more about this feature.

- **SageMaker Edge Manager**: This feature will help you to customize the models for edge devices. The details of this feature can be found here: `https://docs.aws.amazon.com/sagemaker/latest/dg/edge.html`.

- **SageMaker Studio notebooks**: This feature provides support for high-performing notebooks that have integration with Amazon IAM for access management, fast startup times, and so on. To learn more about this feature, please follow this link: `https://docs.aws.amazon.com/sagemaker/latest/dg/whatis.html`.

- **SageMaker Autopilot**: This feature provides support for AutoML in SageMaker. It explores the data, analyzes the problem type, and then builds the optimal model without any client coding required. To learn more about this feature, please follow this link: `https://docs.aws.amazon.com/sagemaker/latest/dg/autopilot-automate-model-development.html`.

- **SageMaker Model Monitor**: This feature helps you to analyze and monitor the model in production and identify whether there is any data drift or deviation in model performance. To learn more about this feature, please follow this link: `https://docs.aws.amazon.com/sagemaker/latest/dg/model-monitor.html`.

- **SageMaker Neo**: This feature helps you to train your model once with the necessary optimizations and then run it everywhere, both in the cloud and on edge devices. Please go to `https://docs.aws.amazon.com/sagemaker/latest/dg/neo.html` to learn more about this feature.

We have mentioned some of the key features of Amazon SageMaker here. There are many other features supported in Amazon SageMaker and the list keeps evolving as time goes on. From the preceding list of features, we can see that Amazon SageMaker provides a large number of tools for the whole ML life cycle. This can be an optimal full stack solution for you to run your data science/ ML projects from end to end.

In the next section, we will create and serve an ML model using Amazon SageMaker.

Using Amazon SageMaker to serve a model

In this section, we will use Amazon SageMaker to serve a model from end to end. You will need an AWS account if you want to follow the examples. Please refer to the *Technical requirements* section to see how to create an AWS account. We will use an XGBoost model created using the same dataset shown here, `https://aws.amazon.com/getting-started/hands-on/build-train-deploy-machine-learning-model-sagemaker/`. We will not discuss the steps to create and train the model here. We will reuse the trained model created in the tutorial at the link.

We will split the exercise into the following subsections for better understanding:

- Creating a notebook in Amazon SageMaker
- Serving the model using Amazon SageMaker

Creating a notebook in Amazon SageMaker

In this subsection, we will create a notebook that can be used to write our code:

1. First of all, let's log in to our AWS account, and we will see the AWS console home page, as in *Figure 15.1*.

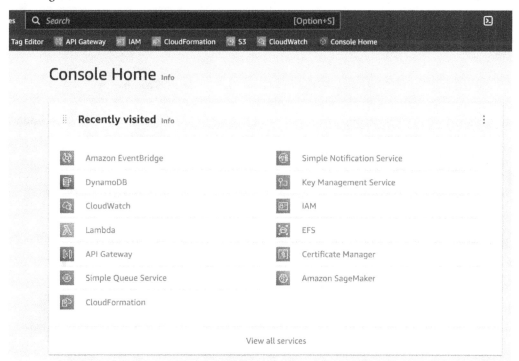

Figure 15.1 – AWS console home page

2. Amazon SageMaker is in the **Recently visited** tools list, as shown in *Figure 15.1*. If you do not see it in the search bar, search for `Amazon SageMaker` and click on the link for **Amazon SageMaker**. This will take you to the Amazon SageMaker home page, as shown in *Figure 15.2*.

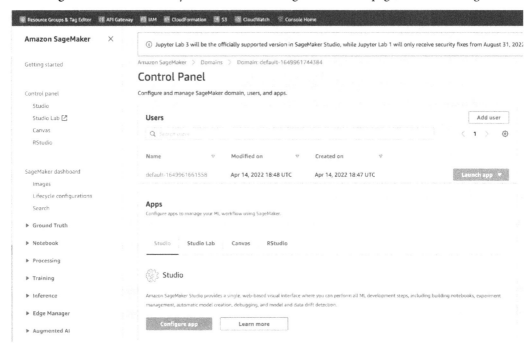

Figure 15.2 – Amazon SageMaker home page in the AWS console

3. Locate the **Notebook** item in the left-hand sidebar shown in *Figure 15.2*. Click **Notebook | Notebook instances** and it will take you to the page listing all the notebook instances, as shown in *Figure 15.3*.

Figure 15.3 – List of all notebook instances accessed by clicking Notebook | Notebook instances

4.   You might see the list shown in *Figure 15.3* is empty. If so, you need to create a notebook. Click on the **Create notebook instance** button, shown in the top-right corner in *Figure 15.3*. After clicking on the button, you should see a window has appeared, as in *Figure 15.4*.

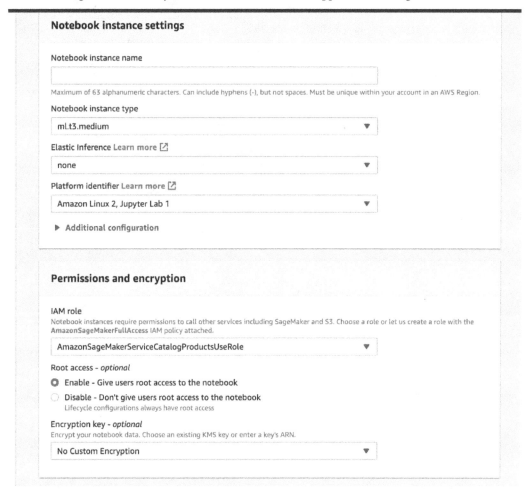

Figure 15.4 – Window for the creation of a notebook instance

5.   Give a name to the notebook in the **Notebook instance name** field. Then, for **Notebook instance type**, select **ml.t2.medium** from the drop-down list.

6.   For the IAM role, from the drop-down list, select **Create a new role**, as shown in *Figure 15.5*.

Figure 15.5 – Options in the drop-down list for the IAM role field

After clicking on the link, you should see the window in *Figure 15.6*.

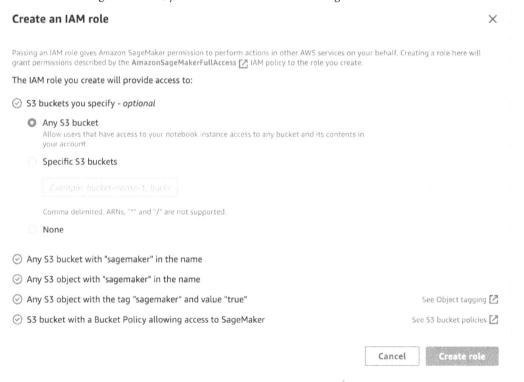

Figure 15.6 – Pop-up window to create an IAM role

7. Then, click on the **Create role** button in the pop-up window.

8. After completing *step 6*, you will be redirected to the page for the creation of a notebook instance. You should see the window in *Figure 15.7*.

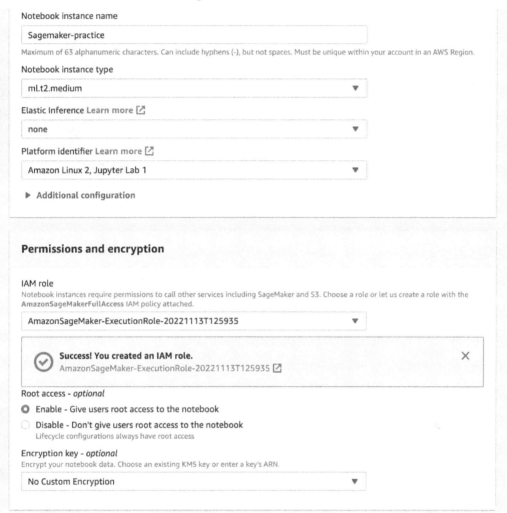

Figure 15.7 – Values of different fields for the creation of a notebook instance

9. Keep all the other fields as their defaults and click on the **Create notebook instance** button at the bottom of the window. Now, it will redirect you to the page where all the notebook instances are listed and you should see the notebook instance that you just created there, as shown in *Figure 15.8*.

Figure 15.8 – Newly created notebook instance is shown in the notebook instances list

Wait for some time (around 2 minutes) until the status of the newly created notebook changes from **Pending** to **InService**, as shown in *Figure 15.9*:

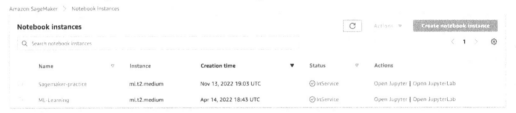

Figure 15.9 – Notebook is ready for access and use with the InService status

10. Now, click on the **Open Jupyter** button, as shown in *Figure 15.9*, and you will be redirected to a new window with the Jupyter Notebook home page opened there, as shown in *Figure 15.10*.

Figure 15.10 – Jupyter Notebook directory where all the notebooks are stored

11. We will click on the **New** drop-down button, as shown in *Figure 15.10*, and we will see a drop-down list, as in *Figure 15.11*.

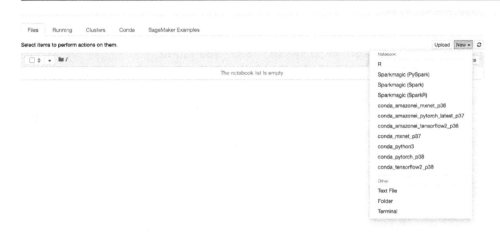

Figure 15.11 – Drop-down menu items for creating a notebook

12. We click on the **conda_python3** option from the dropdown shown in *Figure 15.11*. This will take us to a new window by creating a new notebook where we can write our code. We can name the notebook and it will give us the options to enter cells and run code, as shown in *Figure 15.12*.

Figure 15.12 – Our notebook is ready for writing code

Now, your notebook is ready to create your model.

In the next subsection, we will serve an XGBoost model.

Serving the model using Amazon SageMaker

Amazon SageMaker provides a Docker registry path and other parameters for each of the models it supports. The full list for different regions can be found here: https://docs.aws.amazon.com/sagemaker/latest/dg/sagemaker-algo-docker-registry-paths.html.

We will use the XGBoost algorithm from here: https://docs.aws.amazon.com/sagemaker/latest/dg/ecr-us-east-1.html#xgboost-us-east-1.title. The link gives instructions on how to retrieve the registry path for the model. We follow the steps as shown at https://aws.amazon.com/getting-started/hands-on/build-train-deploy-machine-learning-model-sagemaker/ to train the model, as shown in *step 2* and *step 3* of the page at this link, and then we can serve the model following these steps:

1. To serve the model, we have to run the following command in the notebook to serve the model:

    ```
    xgb_predictor = xgb.deploy(initial_instance_
    count=1,instance_type='ml.m4.xlarge')
    ```

2. Now, let's go to the Amazon SageMaker home page again. Then, click on **Inferences | Endpoint configurations**, as shown in *Figure 15.13*. We can see that an endpoint configuration is already created for the model after we run the command mentioned in *step 1*.

Figure 15.13 – Endpoint configuration list

3. We can click on the configuration and will see a window as in *Figure 15.4*.

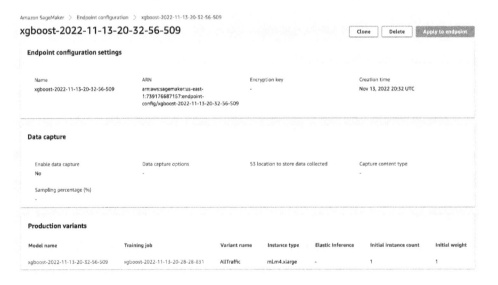

Figure 15.14 – Detailed view of the endpoint configuration

4. Locate the **Apply to endpoint** button and click on that. After clicking it, an endpoint will be created that can be called by clients with appropriate credentials. The endpoint information will be seen in a window as in *Figure 15.5*.

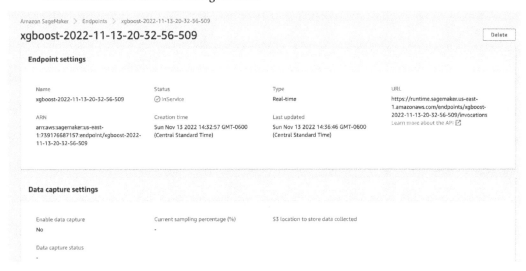

Figure 15.15 – An endpoint is created for inference from the model along with a URL

This URL can be used to send POST **HTTP** requests to this endpoint for inference. To send the request, you will need the appropriate credentials.

5. To be able to send requests from Postman, you need to configure the credentials in the **Authorization** tab in Postman. Select the **AWS Signature** option from the drop-down list, as shown in *Figure 15.16*. Then, provide the access key, secret access key, and session token. To create these tokens, you need to follow the instructions mentioned at this link: https://aws. amazon.com/premiumsupport/knowledge-center/create-access-key.

Figure 15.16 – List of available authorization credentials to select from

Configuring the payload in Postman is also challenging, so it is better to test the endpoint using the AWS-provided Python SDK called boto3. We discuss how we can send requests to the endpoint in *step 6*.

6.   We can create an API client using boto3 and then pass the payload as follows:

```
from sagemaker.serializers import CSVSerializer
client = boto3.client('sagemaker-runtime')
response = client.invoke_endpoint(
    EndpointName="xgboost-2022-11-13-20-32-56-509",
    Body= b'29,2,999,0,1,0,0,1,0,0,0,0,0,0,0,0,0,0,0,1,0,
0,0,0,1,0,0,0,0,0,1,0,0,1,0,0,1,0,0,0,1,0,0,0,0,1,0,0,0,0
,0,0,0,0,1,0,0,1,0',
    ContentType = 'text/csv')
print(response)
print("Prediction is", response['Body'].read().
decode('utf-8'))
```

After running the code, you will see the following output:

```
{'ResponseMetadata': {'RequestId': '936b2e78-97e8-
48b7-9dfe-ca2f7789deb3', 'HTTPStatusCode': 200,
'HTTPHeaders': {'x-amzn-requestid': '936b2e78-97e8-48b7-
9dfe-ca2f7789deb3', 'x-amzn-invoked-production-variant':
'AllTraffic', 'date': 'Sun, 13 Nov 2022 22:42:55 GMT',
'content-type': 'text/csv; charset=utf-8', 'content-
length': '19'}, 'RetryAttempts': 0}, 'ContentType':
'text/csv; charset=utf-8', 'InvokedProductionVariant':
'AllTraffic', 'Body': <botocore.response.StreamingBody
object at 0x7fa8e0cd6fa0>}
Prediction is 0.33817121386528015
```

We notice that we got the prediction as highlighted in the code block. In the preceding code snippet, `EndpointName` is the actual endpoint name that we saw in *Figure 15.15*. As we are running the program in the same account, we do not have to provide any credentials, but if your client program is in a different place, you will also have to provide credentials to send the request. Please refer to this link, `https://docs.aws.amazon.com/IAM/latest/UserGuide/id_credentials_temp_use-resources.html`, if you need to create credentials for API access.

In this section, we have shown how we can serve a model using Amazon SageMaker and create an endpoint for taking inference requests. We have also learned that the AWS security system is built in and cannot be bypassed. In the next section, we will summarize the chapter and conclude.

Summary

In this chapter, we have explored a fully managed cloud solution for serving ML models. You have seen how serving works in Amazon SageMaker, which is a strong representation of a fully managed cloud solution, and you have explored Amazon SageMaker and seen, step by step, how to create a notebook in Amazon SageMaker and how to deploy a model. We have also seen how you can create an endpoint for the model and how you can invoke the endpoint from a client program using `boto3`. This is our last chapter on the tools that we intended to discuss. There are a lot of tools out on the market and a lot more are coming out. I hope, now that you have an idea about serving patterns, you can choose the right tool for you. Amazon SageMaker is a integral ecosystem for ML engineers and data scientists. This chapter only gives an introduction to serving by building a model from scratch using the models from the model registry. There are many other ways to create models, such as using SageMaker Autopilot, SageMaker Studio, and so on. Feel free to explore the world of fully managed cloud solutions to serve ML models.

With this chapter, we conclude our book. We have covered some state-of-the-art patterns for serving ML models and also have explored some relevant tools. I hope this will help you in maintaining the best practice of serving and help fight the stereotype that most ML models do not see the light of day and make it to production. We also encourage you to keep exploring and developing your skills even further beyond what we have discussed in this book.

Index

error rate 74
latency 74
cron expressions 134

D

data drift
reference link 83
data validation 232, 233
DB fiddle
reference link 119
decision path 47
reference link 48
decision tree model
states 59-61
deep learning model 48
bias of hidden layer 52
bias of output layer 53-55
weights from hidden layer to output layer 52
deep neural network 48
deployment graph 259-261
design patterns 16
in software engineering 15, 16
directed acyclic graph (DAG) 200, 201
Distributed Denial of Service (DDoS) 32
Docker
used, for TensorFlow Serving 242-247
DynamicManager 241

E

ensemble model serving
use cases 31
ensemble pattern 218
examples, scenarios 218, 219
techniques, using 219
using, in Ray Serve 261-265

ensemble pattern, techniques
aggregation 222, 223
model selection 225
model update 219-222
responses, combining 225, 226
Error estimation model 196
errors monitoring 77
4XX errors 78-80
5XX errors 81, 82
evaluation period 220

F

F1 score 90, 91
factory pattern 18
feature transformation 234

G

gini 60
gini coefficient 60
gini impurity 60
gini index 60
reference link 60
Google Cloud Monitoring
reference link 75

H

hashing techniques
advantages 131
disadvantages 131

I

ingress deployment 259
intercept 58

Packt.com

Subscribe to our online digital library for full access to over 7,000 books and videos, as well as industry leading tools to help you plan your personal development and advance your career. For more information, please visit our website.

Why subscribe?

- Spend less time learning and more time coding with practical eBooks and Videos from over 4,000 industry professionals

- Improve your learning with Skill Plans built especially for you

- Get a free eBook or video every month

- Fully searchable for easy access to vital information

- Copy and paste, print, and bookmark content

Did you know that Packt offers eBook versions of every book published, with PDF and ePub files available? You can upgrade to the eBook version at packt.com and as a print book customer, you are entitled to a discount on the eBook copy. Get in touch with us at customercare@packtpub.com for more details.

At www.packt.com, you can also read a collection of free technical articles, sign up for a range of free newsletters, and receive exclusive discounts and offers on Packt books and eBooks.

Other Books You May Enjoy

If you enjoyed this book, you may be interested in these other books by Packt:

Production-Ready Applied Deep Learning

Tomasz Palczewski, Jaejun (Brandon) Lee, Lenin Mookiah

ISBN: 9781803243665

- Understand how to develop a deep learning model using PyTorch and TensorFlow
- Convert a proof-of-concept model into a production-ready application
- Discover how to set up a deep learning pipeline in an efficient way using AWS
- Explore different ways to compress a model for various deployment requirements
- Develop Android and iOS applications that run deep learning on mobile devices
- Monitor a system with a deep learning model in production
- Choose the right system architecture for developing and deploying a model

Machine Learning Techniques for Text

Nikos Tsourakis

ISBN: 9781803242385

- Understand fundamental concepts of machine learning for text
- Discover how text data can be represented and build language models
- Perform exploratory data analysis on text corpora
- Use text preprocessing techniques and understand their trade-offs
- Apply dimensionality reduction for visualization and classification
- Incorporate and fine-tune algorithms and models for machine learning
- Evaluate the performance of the implemented systems
- Know the tools for retrieving text data and visualizing the machine learning workflow

Packt is searching for authors like you

If you're interested in becoming an author for Packt, please visit `authors.packtpub.com` and apply today. We have worked with thousands of developers and tech professionals, just like you, to help them share their insight with the global tech community. You can make a general application, apply for a specific hot topic that we are recruiting an author for, or submit your own idea.

Share Your Thoughts

Now you've finished *Machine Learning Model Serving Patterns and Best Practices*, we'd love to hear your thoughts! Scan the QR code below to go straight to the Amazon review page for this book and share your feedback or leave a review on the site that you purchased it from.

https://packt.link/r/1-803-24990-0

Your review is important to us and the tech community and will help us make sure we're delivering excellent quality content.

Download a free PDF copy of this book

Thanks for purchasing this book!

Do you like to read on the go but are unable to carry your print books everywhere? Is your eBook purchase not compatible with the device of your choice?

Don't worry, now with every Packt book you get a DRM-free PDF version of that book at no cost.

Read anywhere, any place, on any device. Search, copy, and paste code from your favorite technical books directly into your application.

The perks don't stop there, you can get exclusive access to discounts, newsletters, and great free content in your inbox daily

Follow these simple steps to get the benefits:

1. Scan the QR code or visit the link below

https://packt.link/free-ebook/9781803249902

2. Submit your proof of purchase
3. That's it! We'll send your free PDF and other benefits to your email directly

www.ingramcontent.com/pod-product-compliance
Lightning Source LLC
Chambersburg PA
CBHW062101050326
40690CB00016B/3161